Evolution Talk

The Who, What, Why, and How behind the Oldest Story Ever Told

Rick Coste

Guilford, Connecticut

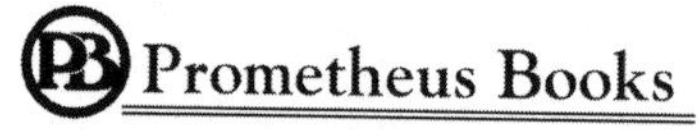

An imprint of Globe Pequot, the trade division of
The Rowman & Littlefield Publishing Group, Inc.
4501 Forbes Blvd., Ste. 200
Lanham, MD 20706
www.rowman.com

Distributed by NATIONAL BOOK NETWORK

British Library Cataloguing in Publication Information Available

Library of Congress Cataloging-in-Publication Data
Name: Coste, Rick, 1967–, author.
Title: Evolution talk: the who, what, why, and how behind the oldest story ever told / Rick Coste.
Description: Lanham, MD: Prometheus, Rowman & Littlefield, [2022] | Includes bibliographical references. | Summary: "Based on the popular podcast of the same name, Evolution Talk reveals how the theory of evolution came to be and how it explains the world around us"—Provided by publisher.
Identifiers: LCCN 2021057514 (print) | LCCN 2021057515 (ebook) | ISBN 9781633888340 (paperback) | ISBN 9781633888357 (epub)
Subjects: LCSH: Evolution (Biology) | Evolution.
Classification: LCC QH366.2.C7347 2022 (print) | LCC QH366.2 (ebook) | DDC 576.8—dc23/eng/20211213
LC record available at https://lccn.loc.gov/2021057514
LC ebook record available at https://lccn.loc.gov/2021057515

∞™ The paper used in this publication meets the minimum requirements of American National Standard for Information Sciences—Permanence of Paper for Printed Library Materials, ANSI/NISO Z39.48-1992

For Mandie.
In a dark and cold universe,
There's a star that is warmer and brighter than all the rest,
And I get to snuggle with her.

Contents

Acknowledgments

The journey that put this book in your hands has been a long and satisfying one. My curiosity about "where it all began" led me to the wonderful books of Richard Dawkins and others before launching my own contribution to the subject with the *Evolution Talk* podcast. The book adaptation led me to Renaissance Literary and Talent. I am happy to say Alan Nevins saw something in this version of the oldest story ever told and turned it over to Jacklyn Saferstein-Hansen. It is no understatement to say that you wouldn't be reading these words or holding this book without Jacklyn. Her care, feedback, and persistence made it possible. She championed *Evolution Talk* until she found it a home. Fortunately, Prometheus Books and Jake Bonar stepped in to provide that home. To Jacklyn, I will be forever grateful.

A big THANK YOU to the life-long learners, university students, and professors who have listened to the *Evolution Talk* podcast. Every email and note of appreciation helped it evolve from a podcast to a book. Who knows what the future may hold for it or what it may transform into next.

And most of all, thank you to my wife, Mandie. Without your endless love and support, I wouldn't be writing the acknowledgments for my book. You're my reason . . . for everything.

Introduction

"Light will be thrown on the origin of man and his history."

—*On the Origin of Species*, Charles Darwin

Published in 1859, *On the Origin of Species* placed Charles Darwin on the front lines of a battle that continues to this day. That battle is the one over our origin story. The origin of the human race. Did we appear on the earth fully formed as we are now, or did we evolve gradually over time from some long-gone ancestor, an ancestor that wasn't human in any way, shape, or form?

Darwin hadn't meant for his book to unleash a firestorm of controversy. What he set out to do was to explain the world and the abundance of species around us. Darwin's theory of evolution by natural selection is perhaps the single most incredible idea anyone has had. Ever. Someone who is genuinely ambitious can apply natural selection theories to anything. In more recent years, the theory has explained how ideas propagate throughout our society and culture. Memes and viral videos are perfect examples of this. Selection theories explain how technology has advanced from a slow, steady rate to the leaps and bounds we have seen over the past 100 years. A beautiful representative of this is the first successful flight of a powered glider by the Wright brothers in 1911. Less than

60 years later, humankind was on the moon. In the span of one human lifetime, an unimaginable amount of achievement occurred. If you trace the evolution of flying machines from the Wright brothers to the space shuttle, you will see a history of small, beneficial adaptations in design. These adaptations allowed our desire for flight to progress from the sands of Kitty Hawk, North Carolina, to the stars.

What does all of this have to do with Darwin's theory of evolution by natural selection? Nothing. And everything.

Without the ability to adapt and survive, we, as a species, would have never seen the light of day.

If you've ever wondered what all the fuss was about or how evolution works, then you've come to the right place. This book examines how Darwin came up with his revolutionary theory and what it means to the life we see around us, but it doesn't stop there. Knowledge of a theory only matters when you can apply it. Using Darwin's theory, we will start from the beginning, the *very* beginning. From the first microorganism to the oak tree outside of your window.

Darwin's ideas were considered dangerous when he first published *On the Origin of Species* more than 160 years ago. Its 150th anniversary was in 2009. Darwin knew his theory would have its detractors, and, as much as it troubled him, he also knew we needed to understand it if we were to move forward.

Darwin wasn't alone, as you will see in the pages ahead. Others had similar thoughts about life on earth and its origins long before Darwin set foot on HMS *Beagle*. These pre-Darwinian thinkers considered how the species we see around us came to be, from the smallest of beetles to the longest-necked giraffe. Using microscopes, dissecting knives, specimen jars, and their own intellect, they considered whether each species was cut from whole cloth. They also wondered if they had transformed into their present states from earlier forms.

No matter what beliefs you bring to the table or your current level of understanding, there is something here for everyone. The theory of evolution by natural selection is not threatening, nor is it difficult to understand. It's so simple you'll wonder why you never thought of it before.

As all stories require a place to start, we'll begin this one in a small English town with a girl named Mary.

Part One

The Pre-Darwinians

Chapter One

The Fossil Finder

There is something fascinating about digging into the earth. Now and then, you may find a piece of the past. A little forgotten glimpse into another person's or thing's life. The more distant that past, the better. As a kid, I loved to play in a sandpit near my house. Looking back, it wasn't a large pit, but to me and my friend Jimmy it was huge. We dug into it for hours, certain that it concealed some forgotten buried treasure. Without the aid of a treasure map, we would think like pirates and pick the spot most likely to give up its secrets. During one of our afternoon digs, I found a flexible gold watchband. How it got there I'll never know. But it fueled our imaginations for more digs and more afternoons. That sandpit became a place full of ankle-spraining holes thanks to us.

That gold watchband turned out to be the only object we'd ever find. Our digs stopped by the time I was nine. I retired my shovel and bought a plaster of Paris kit. Why? To search for and cast Bigfoot tracks, of course. The designs of young minds.

Had I known about a girl whose excavations predated my own by 164 years, I may have kept digging. She hadn't been much older than me when she made her first discovery.

She sells seashells on the seashore
The shells she sells are seashells, I'm sure
So if she sells seashells on the seashore
Then I'm sure she sells seashore shells.

—Terry Sullivan, said to be inspired by Mary Anning

In 1811, or 1812, depending on the account you happen to read, Mary Anning, who was twelve at the time, happened upon a fantastic find. She and her little brother had decided to pass the time digging around the cliffs near their home in Lyme Regis. Lyme Regis is a small coastal town on the English Channel. It's still small today, with a population of not more than four thousand, and it boasts a beautiful set of cliffs made of limestone and shale layers. What can be found in these layers is even more incredible than the vision of the cliffs themselves—as Anning and her brother discovered on that long-ago day.

Anning's family was no stranger to finding odd things in the cliffs. For years they'd pulled the fossilized remains of shells and tiny creatures from the shale and sold them to tourists. Anning's father taught her how to clean the things they found, and the family set up a small table near the town's coach shop to sell their little discoveries. It was the perfect way to supplement the income Anning's father brought in as a cabinet-maker. The cliffs provided more than enough items for the curious and the collectors.

What Anning and her brother had stumbled upon was a skull. A very large skull. Four feet long with jagged teeth. They first thought they'd found a crocodile. The thing was, the area wasn't known for its crocodiles. This was England. Might it be something else, Anning wondered? Unknown creatures weren't a thing yet. The term *dinosaur* wasn't a thing yet. It wouldn't be until the 1840s when biologist and paleontologist Richard Owen coined it, thanks in no small part to Anning's discoveries. Anning and her brother had pulled something out of the cliff, be it a crocodile or not, and decided there must be more. Anning's determination set in, so she kept looking. Her efforts paid off. After more digging, she unearthed the skeleton of a large creature never before seen. Not by human eyes, it was assumed. They sold the skeleton to a local collector who eventually sold it to Charles König at the British Museum. König was so taken by the find that he gave it a name. He called it an ichthyosaurus.

As I mentioned, Anning's family had known there were odd little things to be found in the cliffs. It wasn't until Anning's discovery that anyone suspected the cliffs might contain much bigger "odd little things."

Anning wasn't close to being done, and she continued to dig, her excitement growing with each new find.

Here, hidden in the cliffs and sediment, were bizarre creatures. While some bore recognizable features, like fins and flippers, others were stranger. Imagine Anning's surprise when, in 1821, after years of chipping away at it, she extracted an even more astonishing creature. This turned out to be the partial skeleton of a beast with a long neck. It wasn't a giraffe. She'd clearly found an aquatic creature, so what was it? Anning hadn't a name for it at the time. Eventually, it would be known as the first discovered plesiosaur. By 1823 she had pulled out another one. This time it was an entire skeleton, with a neck so long it contained thirty-five vertebrae.

Anning was quite the artist and became exceptionally good at sketching and describing the things she found. Her ability to identify, sketch, and display the many creatures she unearthed earned her many admirers, as captured in this excerpt from the diary of Lady Harriet Silvester, whose husband had been the Recorder of the City of London.

> *The extraordinary thing in this young woman is that she has made herself so thoroughly acquainted with the science that the moment she finds any bones she knows to what tribe they belong. She fixes the bones on a frame with cement and then makes drawings and has them engraved.*
>
> —September 17, 1824

It wasn't only Anning's sketches and collections that caught people's attention. It was what it meant. With each find, including that of a winged pterosaur she discovered in 1828, questions were raised that begged for answers. What were these creatures, and what happened to them? Had they gone extinct and, if so, why? To add to the mystery, in 1829, Anning also found one of the first known examples of a *Squaloraja* fossil. If you're scratching your head and wondering what that is, so were many scientists at the time. Anning said of her find that it exhibited many of the same characteristics found in both sharks and rays. Could this represent a transitional species? To suggest this was unacceptable. Anning was a Christian, and she knew that her fossils presented a worrying dilemma.

The world was created in seven days, as were all living things. They lived and breathed exactly as they always had. Species didn't go extinct, nor did they evolve into something else. Anning realized, with growing fascination, that the world was much older, and its creatures more varied, than people thought. Whatever these fossils represented, there must be a suitable explanation. Unfortunately, science hadn't one to offer. Not yet. That wouldn't come until years later. Even Georges Cuvier, the respected French naturalist known as the "father of paleontology," failed to recognize what Anning's fossils might mean. So troubled was he by the plesiosaurus, he at first claimed it had to be a fake. It would be years before he changed his mind, eventually calling it "the most amazing creature ever discovered."

Anning's discoveries were eventually to provide the family with a financial boost. Lt Colonel Thomas James Birch, a frequent customer, held an auction of some of his purchases. This was done solely for the Anning family's benefit. By 1826 Anning, then twenty-seven, was able to pool all of her resources together, along with auction's proceeds, to buy a home that served a double purpose as a shop. *Anning's Fossil Depot* was born. It would become a frequent stop for curious fossil hunters and geologists.

If you've been wondering why Anning was so successful when it came to finding fossils, there are two reasons. One is the cliffs themselves. The layers of shale and limestone are especially useful for fossil preservation. The cliffs are also treacherous and prone to landslides. They are especially so when the cold weather rolls in to form ice. The ice serves as natural wedges within the face of the cliffs. In 1833 a landslide almost ended Anning's career. It did, unfortunately, end the bone-finding career of her dog Tray.

Perhaps you will laugh when I say that the death of my old faithful dog has quite upset me, the cliff that fell upon him and killed him in a moment before my eyes, and close to my feet . . . it was but a moment between me and the same fate.

—Letter from Mary Anning to geologist Charlotte Murchison

The second reason for Anning's success was she was simply very good at what she did. She had an eye for it. Whereas I might look at a cliff and see nothing but rocks, Anning would visualize in its craggy contours any number of creatures. A jutting stone might be the snout of a plesiosaur. A flat piece of slate might reveal a flipper. Perhaps, if you were exceptionally lucky and observant, you might find the head of a pterosaur, as Anning did. She pulled out of the cliffs the fossil of a *Dimorphodon macronyx*. This pterosaur flew over Europe some 200 million years ago.

There was a problem with all of this, however. Credit. Anning received very little. When it came to professional publications or official reports on findings, her name was rarely mentioned. Scientists—male scientists—would visit Lyme Regis, stop at her store, and purchase one of her fossils. They'd then bring them home, study them, and present papers on "their" find.

To add to the insult, women were not allowed membership to the Geological Society of London, the oldest of its kind globally. That archaic tradition wasn't obliterated until the doors were opened to women in 1919.

Despite the insult, Anning did maintain friendships with other female scientists, such as Charlotte Murchison, mentioned above. Murchison was a British geologist who collected fossils and spent hours sketching cliffs and other geological features. One can only imagine how many other female geologists might have been remembered by history had the Geological Society opened its doors sooner.

Anning did earn the respect of some of the biologists and paleontologists of her day. England's brightest minds often visited her shop and asked to be shown the cliffs. William Buckland, geologist, paleontologist, and dean of Westminster, was a frequent visitor. Even Richard Owen is said to have gone on fossil hunting excursions along the cliffs with Anning as a guide. Nevertheless, the lack of mention when it came to her efforts and discoveries caused her to be somewhat wary of her visitors at times, and she often found herself questioning the motives of others.

> *The world has used me so unkindly, I fear it has made me suspicious of everyone.*
>
> —MARY ANNING IN A LETTER

And you can't blame her.

In 1835, Anning lost most of her savings to a bad investment. Seeing her plight, William Buckland stepped in to urge the British Government to offer assistance. She was, after all, a valuable asset to the crown. Much to her joy, and possibly surprise, Buckland's efforts were successful, and she was awarded a small annual pension. It was enough to keep things relatively stable for the next few years of her life.

By the mid-1840s, things began to slow down for Anning. It wasn't because the cliffs' abundance of fossils was drying up, not even close, but Anning's health had taken a turn for the worse. Anning was diagnosed with breast cancer, a disease she fought against bravely. That battle ended on March 9, 1847. She was forty-seven years old.

The fossils Anning collected are still out there. You can see them in any number of museums. The Natural History Museum in London has displayed her ichthyosaur, pterosaur, and plesiosaur. You can also find some of her discoveries at Oxford University. Her fossils live on, yet Anning remains virtually unknown to the general public. A few years after her death, the Geological Society of London, the same one that would not accept her as a member at the time, gifted a stained glass window to the church in Lyme Regis in Anning's honor. It is still there today.

In 2010, the Royal Society listed Mary Anning as one of the ten most influential women in the history of science.

Mary Anning's discoveries around the cliffs of Lyme Regis have survived the ravages of time. She would undoubtedly be happy to know her efforts helped give birth to an appreciation of the creatures that made their home on this planet long before we showed up. These creatures boggled the minds of those who saw their remains and would one day grace the screens of popular culture and *Jurassic Park* films. One can only imagine the fascination those first few shell discoveries held for Anning as she learned to clean them next to her father. Those fossilized shells were of a different world, and each told a different story. A secret story. I like to think it was the pursuit of those stories that brought Anning back to the cliffs day after day. It was all about the untold tales those fossils revealed. The strange glimpses into another world that time had forgotten. It explains why she spent years painstakingly and meticulously digging away at the cliff to release the plesiosaur from its stony grave. It was the stories of the cliff that attracted Anning. She learned from them. And, in turn, we learned from her.

Accounts of strange creatures that once walked, swam, or flew over the earth continue to capture our imagination. But that's not all they do. For others, like Anning, they reveal a mystery. Where are these creatures now, and do they still exist, not in their original forms but hidden in the animals we share the world with today? And what does it mean for us as a species? Anning's fossils tell a story, one we are still learning more about with each passing decade.

That story explains how the world came to be populated by the life we see around us, and it's a very old story.

Long before Mary Anning and her brother found the ichthyosaur's skull, those questions were raised, and some answers were offered by other great minds.

Let's look at a few.

Chapter Two

Aristotle and Design

For in all natural things there is something marvelous.

—Aristotle *Posterior Analytics*

Over 2,300 years ago, Aristotle (384–322 BCE) considered the animals. Within each animal, the "Father of Zoology" thought, there is something special. Something worth acknowledging and studying. Each animal has a story to tell. To understand our own story, we must first understand the creatures we share it with. There are secrets to uncover and great truths to unveil, and they can be found in the smallest worm. We only have to look and to ask the right questions. If the questions aren't entirely obvious yet, they will be. Nature will reveal them to us when we turn our attention to finding the truth. And truth, most philosophers will tell you, isn't subjective. A thing is true, or it is not.

Aristotle sought to uncover the truth. About everything. If he had his way, no stone would be left unturned. Literally. The Creator had salted the world with the grains of truth, and it was up to us to find them. The answers were all around us.

For twenty years, Aristotle attended Plato's Academy in Athens and listened to his great teacher's ideas about the world. One of Plato's favorite topics was his concept of forms. Forms, according to Plato, are immutable ideals that we recognize on a subconscious level and that exist in a different realm. When we see a horse, we know it to be a horse.

But why? How do we know the animal in front of us is a horse? We've given it a label, an identifier taught to us at an early age, but how can we recognize a horse every time we see one? According to Plato, what we recognize, on an intuitive level, is the ideal form of a horse. This perfect form is unreachable by the senses. It is imperfectly represented by the material form we see in front of us. I can draw a circle, but it will never be a circle without fault. I may take my time to draw it, but there will be flaws, however small and invisible they might be. My carefully drawn circle symbolizes the perfect circle in Plato's unreachable realm of ideal forms. You can find forms representing the ideal dog, the ideal chair, and the ideal human in this realm. Plato would tell you forms do not change from one to another, at least as far as species are concerned. There can be no such thing as "transitional" ideal forms. A horse is a horse is a horse.

Aristotle disagreed with his teacher. He refused to contemplate a world that can never be reached by the senses. For him, all that mattered *were* the senses. It's all we have. To speak of the ideal sea sponge is one thing. To touch, study, and examine a real sea sponge with your hands, to see it with your eyes, is another thing altogether. What we can see and touch tells us more about why we are here than an unattainable realm of perfect forms ever can.

Nature fascinated Aristotle. So much that he wrapped himself in it. He awoke at sunrise, hastily ate breakfast, went outside, and plunged his hands into the soil and the sea. He wished to see for himself the marvelous creatures the Creator had given us as companions on this world. Understanding them would bring him closer to the truth. Aristotle sought to classify all animals, including man, by placing them upon a complexity ladder. At the very top of this ladder is the Creator, the One who started it all. The Unmoved Mover. At the very bottom of the ladder are rocks. Situated on the rungs between the top and the bottom are plants, animals, and man in order of complexity. Each is classified according to how intricate they are and the function or purpose they serve.

Without purpose, nothing would exist. Rocks, Aristotle would argue, have a purpose. Without them, we would have nothing to stand upon. There would be no earth, no moon, and no heavenly objects. Rocks and minerals form the base of it all. They are the pedestal upon which everything else is displayed. Plants are more complex than rocks but less complicated than animals. Plants do not enjoy motion. They are not mobile, cannot move around, and do not employ the type of digestive system animals have. Plants also do not have blood. Aristotle was quick to point out that bloodless creatures are less complicated than creatures with blood in their veins. Man, according to Aristotle, sits above it all and is positioned right below the Creator. Because of our brains, and our ability to perceive and contemplate the world, we are set apart from everything else. For each unique property, there is a purpose. Our brain allows us to question why we are. It also allows us to discern answers from the truths scattered around us.

We have the Creator, the One at the top of the ladder, to thank for all this. Aristotle proposed a teleological view for the design and presence of structure in nature. Teleological means it is driven by purpose and purpose alone. Nothing is by chance. There is utility and motive in nature.

To find utility and purpose, we only need to look at our reflections in the water. We have eyes so we can see. Our ears enable us to hear. Our hair? It is there to keep our heads warm.

If only the world were that simple.

What Aristotle missed, or what he refused to see, is that we do not need purpose to explain complexity. It has nothing to do with intention. It's a blind experiment of trial and error. An experiment without an experimenter. This blind experiment accounts for the diversity of species. A seed is planted, and it sprouts off into many branches. The great, tangled branches of life we see today came from a single seed. We are no higher on the chain than any other creature. We are simply different.

Poetry is written in this way. From a single idea springs beautiful prose composed of words available to the poet. No two poems are alike. They evolve as well. A poet's view of the world is perhaps the most beautiful view of all.

In the next chapter, we will meet such a poet.

Chapter Three

Evolution's Poet

Poems come in many shapes and sizes. Not so different from life here on our little planet. Take the poem *On the Nature of Things,* written by the Roman poet Titus Lucretius in the first century BCE. It contained 7,400 lines of verse that Lucretius split into six self-contained books. I say self-contained because each book focused on a different aspect of nature. Still, when read in its entirety, section by section, a clear argument emerges. In *On the Nature of Things*, Lucretius seeks to explain everything. You can look at it as one exceptionally long argument. He covers it all, from the tiniest particles of matter and how they move to the nature of time, space, consciousness, mortality, and the arrival of life. The poem is a well-thought-out manifesto for materialism. Materialism offers the viewpoint that everything follows the rules of physics in nature. That is all. There is no wizard behind the curtain or a master designer hidden from view (sorry, Aristotle). Without materialism, many claim, everything would be smoke and mirrors.

We don't know a lot about Lucretius's life. Other than being a poet, all we can say with any level of certainty is that he lived in the first century BC and died around 55 BC after an apparent suicide at age forty-four. We don't know for a fact he committed suicide. The claim that Lucretius died by his own hand came by way of Christian theologian Eusebius Hieronymus, also known as Saint Jerome. Jerome wrote that Lucretius went insane after drinking a love potion. The resulting madness led him to take his own life. We can never know whether this is true.

Jerome's account of Lucretius's life was presented four centuries *after* he died. That's not a very reliable source considering nothing else was written between the years of the poet's death and Jerome's sad account.

What else do we know? We know Lucretius was a follower of Epicurus. Epicurus was a Greek philosopher who lived during Aristotle's time and rejected the idea that God was the one responsible for everything. We also know Lucretius wrote an amazing piece of work—the poem mentioned above called *On the Nature of Things*. Lastly, we know Lucretius died before completing it.

That's about it. Lucretius was born. He lived in ancient Rome. He may or may not have been a lovesick poet, and he worked on his magnum opus until the day he shed his material form. You might wonder how we came by this poem if Lucretius died before completing his final draft. Fortunately, there were copies made, and these floated around ancient Rome for a while, until they were lost.

By pure happenstance, Roman scholar Poggio Bracciolini found them again in 1417. His friend Niccolò de Niccoli made a copy, and we have that copy today. It's in the Laurentian Library in Florence. The very first edition of the work was printed in 1473, and it has remained in print ever since. The third president of the United States, Thomas Jefferson, himself a materialist, owned a few copies in different translations.

Why are we talking about a philosophical work written by a poet over two thousand years ago in a book about evolution? Especially a poet we know very little about?

It's because of Book Five.

In Book Five, Lucretius sets out to describe how life here on earth came to be. He starts with a contemplation of the earth, which he refers to as the "Mother." The moment life originated on earth, it happened once, and it will happen no more. Lucretius compares the end of his origin story with that of a woman who has reached menopause and can no longer bear children.

Ever the materialist, Lucretius tells us life appeared in nature because of physical laws and nothing else. Think of the "Mother" not as a designer but as an accidental laboratory.

In the next set of verses, he sets the stage to explain why some creatures born of the earth, in the beginning, are no longer with us.

And in the ages after monsters died,
Perforce there perished many a stock, unable
By propagation to forge a progeny.
For whatsoever creatures thou beholdest
Breathing the breath of life, the same have been
Even from their earliest age preserved alive
By cunning, or by valour, or at least
By speed of foot or wing. And many a stock
Remaineth yet, because of use to man . . .

Did you catch that?

What Lucretius casually threw out there is a mechanism in nature that explains why some creatures are no longer with us. They were not fit to survive.

Remember our friend Aristotle and his ideas about purpose? Aristotle attributed each step, or each modification, to an end goal or a future state. Mankind represents the fruition of those steps, and Aristotle placed us a couple of ladder-rungs below the Creator. Lucretius disagreed. There was no Creator, nor was there a purpose.

But origin of tongue came long before
Discourse of words, and ears created were
Much earlier than any sound was heard;
And all the members, so meseems, were there
Before they got their use: and therefore, they
Could not be gendered for the sake of use.

Here, Lucretius tells the reader that our ears and tongues were formed long before they had a use. Ears served a purpose at one time, or an earlier form of an ear did, but it was co-opted for hearing. We know this today. Evolutionary and comparative biology have identified many body parts that have seen their uses change over time and between species. This is called exaptation instead of adaptation. There are instances of this with our anatomy and the anatomy of other species with whom we share a common ancestor. One example is our middle finger. You might say we have fingers for the sole purpose of picking things up. Yes, we use them for that now, but our fingers evolved into their shape because they benefited our ancestors. So did a horse's hoof. The horse's hoof *is* our middle finger. Comparative biology has shown us that the horse's hoof still contains remnants of the missing fingers, and that what we call a hoof is actually an enlarged middle finger. The bone structure of a horse's forelimb is the same as the bone structure of our forearm. The bones are all there. They have been molded into different shapes over billions of years. It is as Lucretius said. It was all there before they reached their present state of use.

Another modification is our inner ear bones. Those same bones exist in reptiles, but they are part of their jaw. Over time those bones moved back to where our ears are and now have a different use than they originally did.

On The Nature of Things lays out the creation of the universe from a chaos of matter to the formation of our earth and the appearance of living and breathing creatures through natural processes. Those creatures fought for survival. The ones better able to adapt to their environments survived. Those not so good at it perished. Body parts were co-opted and modified by the same natural processes that formed the universe. Shapes changed over time, and those changes gave some creatures an advantage over their competitors.

Evolve and adapt. Like thoughts and ideas. They evolved over grand expanses of time. Ideas survive because there is truth to them.

In the words of Lucretius:

> *No fact is so obvious that it does not at first produce wonder,*
> *Nor so wonderful that it does not eventually yield to belief.*

It would be a few hundred years before another writer saw fit to pick up where Lucretius left off. That man was Benoît de Maillet. Let's see what he had to say.

Chapter Four

Of Mermaids and Men

Great ideas aren't conceived in a vacuum. Think of them like jigsaw puzzles with imperfect pieces. Some fit together better than others. The pieces may appear to be made for each other and are left that way until someone realizes they were forced together. A lot of ideas and arguments are later discovered to have been formed in this manner. It's lazy puzzle making. Some minds see the picture before all the pieces are in place and then try to tackle it. No matter your approach, the first step is to lift the lid off the box and dump the pieces on the table.

A puzzle-master might tell you the best place to start is with the edges. Those straight lines begin the story. They form the outline. Once we have an outline, it is time for the hard part. The bigger the puzzle, the more complex. The more complex, the greater the chance the eager puzzle-solver will only get so far before quitting—leaving it there for someone else to complete it.

One whose contribution to the puzzle is often forgotten is Benoît de Maillet. Maillet's contribution to the puzzle is a book. The book is called *Telliamed.*

In 1692, Maillet lived in Cairo, where he served as a French consul. By all accounts, he did his job well. He was curious about many things. This inquisitiveness took him into areas most men of his stature would have stayed away from. He had duties to perform, which he did during the day. At night he would retreat to his quarters and write. There were letters to be written, letters bound for France, as was his duty, but there were also pages of a book. Pages he preferred to keep hidden during the day.

Why?

It's because those pages contained observations and ideas some might say were dangerous. The kind of ideas that would have caused Maillet to be stripped of his position and sent back to France.

What was so risky about Maillet's secret project? Benoît de Maillet felt he had figured out the origins of life. He didn't stop there, either. He also deduced, or so he thought, how the earth came to be populated.

Maillet started his book in 1697. The ideas that pushed him into completing the first few pages surprised even himself. His curiosity had driven him to those ideas. It led him to books by ancient philosophers. In them, he found novel ways of looking at life and the world around him. The earth itself, the continents and landmasses, appeared to shift and mold themselves into different shapes and forms. Not in any perceptible way either. He came across old and dusty texts that described sea journeys between Alexandria and the islands of Pharos. What struck him was that these journeys were unnecessary. In his time, these landmasses were connected. Why make the trip by sea when you could hop on a camel and journey there by land?

There were other examples. He read about and saw for himself giant rings embedded in stone in Memphis, Egypt, once used to moor great boats. He marveled at this and imagined great ships moored at docks that no longer existed, their long ropes tied to these massive iron rings. The problem was that Memphis was inland. There were no shores for boats to arrive at. There was only sand.

Observations like this led Maillet to conclude that the earth was extremely old. Not thousands of years old, as many believed, but billions of years old. Old enough for the land to change and transform into different shapes. Shapes his ancestors wouldn't recognize. The changes wouldn't be noticeable during anyone's lifetime, but as generations came and went, the changes were there.

In 1697, at the time Maillet wrote his book, he was forty-one years old. If you had told him he would work on his book for the next forty years, he wouldn't have believed you. Nor would he have wanted to. He didn't finish it until 1735. To say he finished it is a little misleading. What he had in 1735 were chests full of pages in need of a massive number of edits.

Why had he worked on his book for so long? Maillet wanted his ideas to be airtight. He was a perfectionist. His job as French consul meant he had to have an eye for detail and the ability to discern truth in chaos.

Maillet believed that life, all life, came from the sea. Not only did it come from the sea, but it continued to evolve into different species as those species spread out to encounter different environments.

> *Who can doubt, that from the volatile fish sprung our birds, which raise themselves in the air; and that from those which creep in the sea, arose our terrestrial animals which have neither a disposition to fly, nor the art of raising themselves above the earth?*
>
> —Benoît de Maillet, ***The Telliamed***

To present these ideas publicly would be hazardous. Maillet was a respectable man tasked with an important job. To publish such a book would ruin him, his career, and his reputation. Everything he'd worked so hard for would be lost. There was the question of publishing it under a different name when he felt it was ready. He immediately discarded this notion. Maillet was too proud to publish his life's work anonymously. It was his masterpiece, after all.

And then it hit him.

Why not present his work as a work of fiction instead? There were plenty of fictional tales in bookseller's stalls that spoke of fantastic ideas and voyages. If he were to claim birds sprung from fish, they would brand him a heretic, but if a character in a story were to say the same thing, he would have nothing to worry about. He could always roll his eyes, shrug his shoulders, and say, "My character is crazy, is he not?" So that's precisely what Maillet did. He created a character named Telliamed. Telliamed became his mouthpiece.

If you think Telliamed is an odd name, it is. It's Maillet's own last name spelled backward.

In his book, Telliamed is an Indian sage gifted with great and strange wisdom, which he imparts to the story's narrator. Telliamed's insight includes the idea that life on earth originated in the sea. One has only to look at that life closely to see the evidence. It's all there in the elements.

If we pass metals, earth, wood, plants, animates or inanimate bodies, whatever the earth includes or produces, and in short, fresh water, through an alembic, we find salt in each of them, and discover the vestiges of the sea, to which all things owe their origin.

Like Lucretius had said in his poetic works, there was no Creator in Maillet's tale. Everything was chance. From primitive creatures sprang life—even man.

There is in all men an indelible mark, that they draw their origin from the sea. In a word, consider their skin with one of our lately invented microscopes, which magnify a grain of sand to the bulk of an ostrich's egg, and you will find it all covered with small scales like those of a carp.

Instead of slowing down or trying to organize his work into something a little more manageable, Maillet forged on. He needed to capture every idea before it slipped away. He wanted people to accept what he said, even if he presented it as an Indian sage's fictional utterances.

As time passed, Maillet grew worried. The book had become too big. It was so immense, so expansive that he couldn't find a way out of it. His worry gave way to panic. He wasn't a young man anymore. If something happened to him before he completed his work, it would be lost forever.

In desperation, he turned to fellow French author Bernard Le Bovier de Fontenelle for advice. Fontanelle told Maillet he needed to add more to his book and that it needed more substantial evidence. Not exactly what Maillet wanted to hear. But, according to Fontanelle, if Maillet expanded on some of the ideas found in his book, it would be unstoppable. It would become a sensation. There was not a person in the world who didn't want to know the origins of life. If those origins contradicted what the church had said for centuries, all for the better. The public loved controversy.

At Fontanelle's bidding, Maillet poured himself back into his work. He had to show that organisms adapted to new environments and developed into different species over time. In plants, he saw seaweed, and in birds, he saw flying fish. But what of man? Where was man's connection to the sea? If flying fish became birds, and fish crawled out of the ocean to become animals, what accounted for man himself? Maillet needed a transitional species. One that stood between fish and man.

"Ah," he thought. "Mermaids."

In stories of mermaid sightings, Maillet felt he had the final piece to the puzzle. Mermaids, and mermen, were the transitional species between fish and man he'd been looking for. To fortify his argument, he collected stories of sightings. Some were reasonably well documented, and he relied on these stories to support his case. There were tales of strange men and women who emerged from the sea but found themselves unable to cope on land. There were other tales of beastly, bearded men who burst forth from the waves with fins instead of legs. Maillet felt he might have all the pieces he needed to finally put his puzzle together.

To help provide his book with some semblance of structure, he handed it over to fellow French writer Jean Baptiste de Mascrier. Mascrier would serve as editor to Maillet's monumental work. It wouldn't be a simple job. To further complicate matters, Mascrier had to deal with Maillet himself. Maillet couldn't stop writing. On any given day, Mascrier received new material to add to the manuscript. The additions, and subsequent edits, were endless.

And then the letters to Mascrier stopped. When Mascrier sought to find out why Maillet had stopped writing, he encountered the sad truth.

Benoît de Maillet had died at the age of eighty-two.

With the storm of edits and requests to add material behind him, Mascrier breathed a sigh of relief and took his time. A lot of time. It wouldn't be until 1748, ten years after Maillet had died, that Mascrier believed the book ready for publication. The sad part is, not only had Maillet died before seeing his life's work reach the public's eyes, but there were portions of the book he wouldn't have recognized. Mascrier had taken liberties Maillet would have been horrified to read.

To make Maillet's book more palatable, even as a work of fiction, Mascrier had decided to make a few adjustments. Maillet wasn't around to stop him or ask for permission, so Mascrier did what he thought best. One of his changes concerned the role of God. Maillet's book had no place for a Creator. Blind chance and random processes produced the earth and life itself. Catering to the watchful eye of the church, Mascrier introduced the hand of God. He also structured the book into six sections to represent the six days of creation. Maillet would have been appalled and heartbroken.

He would have also been pleased with its reception.

The book was a sensation, *and* it was scandalous. It had precisely the impact Maillet had hoped for.

It would be 230 years after his death before Maillet's original manuscript was painstakingly pieced together by scholars. Mermaids and all.

Unbeknown to Mascrier, while he was busy forcing together the puzzle pieces of Maillet's massive work, another Frenchman, this one a mathematician, pondered over a puzzle of his own.

Chapter Five

Saint-Malo's Mathematician

The French city of Saint-Malo has an exciting past. You could call it a checkered past. I like to think of it as a colorful one. In 1590, it declared itself an independent republic, one that was autonomous from France. Its sovereignty lasted for three years. Saint-Malo sits on the northern coast of France along the English Channel. Enormous walls surround the city itself. In its early days, it was home to the Corsairs. The Corsairs were legalized pirates as far as the French crown was concerned. If you were an enemy of France and sailing the open waters, you were fair game for the Corsairs. They auctioned the spoils off, and the government entitled the Corsairs to keep a good portion of the proceeds. They were very well off.

It was into this world of privateers and pride that Pierre-Louis Moreau de Maupertuis was born in 1698. Maupertuis's family were wealthy and socially connected. His family's status allowed young Maupertuis to enjoy the finer things in life. For him, the most pleasing endeavor he could indulge himself in was mathematics. Math, for Maupertuis, was as essential to life as the air we breathe. It moved him like no other pursuit. He threw himself into and championed the work of Isaac Newton. The latter had died less than a decade before Maupertuis was born. Since he was an Englishman, Newton's theories were continuously called into question in France. Maupertuis's loyalties landed on Newton's side. He joined the debate to swing the pendulum of favor back to his idol.

One of Maupertuis's early mathematical endeavors, and eventual victories, was to argue, based on Newtonian physics, that the earth was oblate, meaning that it is a somewhat flattened sphere. Imagine taking a semi-inflated basketball and pressing down on it from the top and bottom. The opposition believed that the earth was prolate. Take the same basketball and hold it between your two hands, and press inward from the sides.

The "Oblates" won. Maupertuis was right, and they celebrated him for it.

Maupertuis's math expertise landed him a position as the director of the Academy of Sciences in Paris. This honor eventually led to his selection as the president of the Berlin Academy. That was an odd position for him to hold, since Maupertuis himself did not speak any German whatsoever. The situation became precarious when France and Germany went to war between the years 1756 and 1763. History knows it as the Seven Years' War, which was essentially one of the first world wars.

Math wasn't the only thing that fascinated Maupertuis. Life did. Or, to be more specific, the origin and evolution of life. Math may hold the keys to unlock the secrets of the universe, but the keys didn't quite fit when it came to the secrets of life. Math was perfect for a lock that required a single key, even an elaborately chiseled key, but life's mysteries were behind a combination lock with a multitude of dials that boasted hidden numbers and secret codes.

At the time, the subject of life and its beginnings fell to natural theologians to explain, which they did in no uncertain terms. The natural theologians of Maupertuis's day pointed to the Creator. God was the grand designer behind it all. A frog was a frog and had always been a frog. Any resemblance to another animal or species was simply the case of a design being reused and tweaked. God gave many animals four legs and reserved his bipedal designs for those animals of higher intellects. His crowning achievement was us.

Maupertuis didn't buy any of it. There were too many inconsistencies in logic as far as that argument was concerned. He was a mathematician, after all. If there was a Creator, finding the keys to his work involved careful study of the facts and an examination of the natural world with critical eyes.

Like any other problem he'd faced, he broke it down into a series of premises.

Chance, one would say, produced an innumerable multitude of individuals.

This was Maupertuis's first premise. One that he introduced in his 1745 work, *Venus Physique*. It was chance that started it all. Life simply popped into existence. It didn't require a Creator per se. It just happened. Organic matter was self-organizing. From that moment, when those very first life forms crawled across the planet's surface, things happened—the most incredible things.

Maupertuis considered those little life forms . . .

A small number found themselves constructed in such a manner that the parts of the animal were able to satisfy its needs.

Those animals and plants were the lucky ones. Others weren't so lucky.

In another infinitely greater number, there was neither fitness nor order.

Those poor individuals, the ones hampered by a lack of fitness or order, were pushed aside by those better able to satisfy their own needs. What happened to those less fortunate individuals?

All of these latter have perished. Animals lacking a mouth could not live; others lacking reproductive organs could not perpetuate themselves.

Maupertuis wanted to understand how organisms changed. Once they'd changed, they would either survive long enough to reproduce, or they wouldn't. In a changing environment, what was it that prompted the changes? Perhaps traits were passed down from parents to child? Were features mixed together to create new variations in form? To prove this, Maupertuis pointed to individuals who'd inherited an extra finger. We know this in today's medical terms as polydactyly. Having offspring exhibit this anomaly meant a particle was passed on that had to account for this specific oddity. It wasn't the result of blending. It resulted from a unique and inheritable feature encased in a particle. What precisely that particle was remained to be seen.

In a real show of forward-thinking, Maupertuis considered the entire history of the planet. At the same time, he wrote about the transmutation and propagation of species. From its spontaneous and uninspired beginnings, life marched forward. Many species were left behind. In our limited and hampered ability to see into the past, we will be forever unaware of them.

The species we see today are but the smallest part of what blind destiny has produced.

On July 27, 1759, at the age of sixty, Pierre-Louis Moreau de Maupertuis joined the organisms that had passed into the shadows of time before him. He left behind no wife to mourn his passing nor children to carry on his genes. He left us with his words, and sometimes that is enough to be remembered by, as we will see in the next chapter. One has to wonder if Maupertuis had the pleasure to read a little book published in 1749 titled *Letter on the Blind for the Use of Those Who See.*

Who wrote this book, and why do we care? Let's take a look.

Chapter Six

Diderot's Dream

Can a book be provocative enough to hinder society's growth to the extent that writing, or even publishing such a book, would be a crime? One punishable by a few years in prison or, in extreme cases, even death?

There was a time when this was the case. There have been authors who were hounded by authorities because of their books.

Denis Diderot was one such author.

Diderot, a Frenchman, was born in 1713. For all intents and purposes, he conducted himself as would any respectable young man of modest means. He studied philosophy and considered entering the clergy. These were heady times for young Diderot. The "Age of Enlightenment" revolutionized how we thought about the world. Ideas about God and man were abundant, especially when those ideas concerned the *evidence* for God. According to Diderot and many others at the time, God's work could be found in nature. These words would be echoed by William Paley a century later, when the clergyman compared the intricate workings of a watch to the complexity he found in the natural world.

Diderot would eventually obtain a degree in philosophy. Instead of lending his voice to the argument that the apparent designs in nature were evidence of God, his studies took him into a vastly unexpected direction. The more Diderot read, the more he questioned his previous views on humankind's place in nature and how human beings even came to have a place.

Diderot devoured the written word. His appetite for it knew no bounds, and he was greedy for ideas and books. His passion grew when those books took him to places others feared to tread. A vast black market existed for these books, and Diderot tapped into it.

Subversive books popped up everywhere. One of the more scandalous tomes was written by the French philosopher Julien Offray de La Mettrie. The book was called *L'homme machine* (*Man a Machine*), and in it, La Mettrie compared humans, and all of their inner workings, desires, and actions, to an automaton. We, like mindless golems, operated without guidance and never strayed from physical mechanics. La Mettrie claimed free will didn't exist, nor was there a guiding spirit or soul behind our movements, decisions, or emotions. In other words, there was no ghost in the machine. There were only the mechanics of an eye and a brain. Any thoughts produced by the eye and the brain's interaction were in themselves the result of physical processes.

Another book that caught Diderot's attention was Maillet's *Telliamed.* Maillet's notion that life on earth originated in the sea and evolved from simpler forms fascinated young Diderot. Maillet's book even claimed that man himself emerged from the sea. How else can one explain the proliferation of stories about mermaids and mermen who occasionally popped their heads out of the waves to survey the land?

Diderot was mesmerized. Could these authors be onto something the church, and even the state, wanted to keep hidden from view? What else could explain the persecution these authors feared and fought against? France had a long history of censorship, and no time was worse than the time Diderot grew up in. The government tasked the director of Book Trade to approve all publications and prosecute authors who illegally published books of "heresy, sedition, and personal libel."

Diderot set his concerns aside and wrote. He had a voice to add to the wave of new thoughts that swept through France. Unlike Maillet, Diderot felt it best to publish his work anonymously. He might be proud of his ideas, but he would be less so to find himself behind bars. One of his first anonymous productions was a book published in 1749 called *Letter on the Blind for the Use of Those Who See.* In it, Diderot wrote against the providence of God. He hinted that God wasn't needed when it came to nature. Species evolved, and mankind was nothing more than thinking matter.

If he thought publishing his works anonymously would protect him from the eyes of the law, he was unfortunately mistaken. The police identified him as the author of *Letter on the Blind*. As a result, Diderot was unceremoniously thrown into prison.

Upon his release, Diderot brushed himself off, promised not to do it again, and proceeded to do it again.

His time in prison hadn't been without benefit. Diderot had kept busy. He planned an alternative approach to getting his ideas out into the world. Boldly stating where he stood on certain subjects wasn't the right way to do it. He had to be smart about it. One way to do so would be to cloak his thoughts in "what if?" scenarios. He could pose these scenarios as questions and then swiftly refute them. The beauty of this is his refutations didn't need to be conclusive. They could be as strong or as weak as he wanted them to be. The rebuttals weren't important; the "what if?" scenarios were.

We can find a perfect example of this in his work *Thoughts on the Interpretation of Nature*, published in 1754.

Diderot asked, may it not be that man . . .

will finally disappear from nature forever, or rather, will continue to exist, but in a form and with faculties wholly unlike those which characterize him in this moment of time?

He followed this up with a brief answer: *But religion spares us many wanderings and much labor.*

And, just like that, he had his method of writing. He could talk about anything he liked, even the gradual evolution of man into a different species, and then refute it by stating that religion spares us from these evil thoughts. It was, in a word, brilliant.

Diderot made a name for himself within Paris's intellectual circles, many of whom had ideas and thoughts similar to his own. They were forever under the watchful eye of the Parisian police as well. The Enlightenment may have been illuminating the dark corners of human knowledge, but many were willing to extinguish that light. It was under these conditions that Diderot put together another subversive work known as

D'Alembert's Dream. He continued to utilize the "present and dispute" method of advancing the ideas he wished to convey, but in *D'Alembert's Dream*, Diderot did this by organizing the work into three separate conversations. The first conversation was between Diderot and D'Alembert. The remaining discussions involved Julie de Lespinasse and a character named Doctor Bordeu. One thing that made these conversations stand out was that D'Alembert and Lespinasse were actual people whom Diderot knew intimately. D'Alembert was a mathematician and philosopher who had worked with Diderot before, and Lespinasse moved within the same intellectual circles. Neither D'Alembert nor Lespinasse were pleased to find themselves used as protagonists in his work. Especially one certain to be viewed with displeasure by the director of Book Trade. The character of Doctor Bordeu was fictitious and served as a stand-in for Diderot himself. Bordeu was a suitable mouthpiece when the ideas presented proved too controversial. To say they were controversial is an understatement. What Diderot delved into was the very thing that Benoît de Maillet had started almost a century before.

To put it simply . . .

Every animal is more or less a human being, every mineral is more or less a plant, and every plant is more or less an animal. There is nothing fixed in nature.

Diderot proposed that species evolved into the forms they were in now. They had changed over time. His message was clear.

That species transformed into other species over significant periods was contrary to everything the church taught. The clergy knew how life originated and who had started it all. God did. In Diderot's work, species resulted from gradual transformations and permutations, and humanity wasn't excluded from this process. We were just another part of it, albeit a significant part. As human beings, we evolved the ability to question our origins in an effort to understand our place in nature. Publishing books that championed these theories contradicted the teachings of the church. By not using his own name as a character in *D'Alembert's Dream*, Diderot didn't have to own what he said.

Diderot claimed that not only did species evolve, but everything is comprised of the same cells. The differences we see are because these cells take different forms. Like a box of Legos. We can make the blocks into virtually anything the mind can dream up. Nature works the same way. It is in a constant state of flux—everything changes.

Diderot also proposed something that would be resurrected centuries later by biologist Stephen Jay Gould in his book *Wonderful Life*. "Wind back the tape of life to the early days of the Burgess Shale; let it play again from an identical starting point, and the chance becomes vanishingly small that anything like human intelligence would grace the replay."

Diderot's words were eerily similar:

Once the sun goes out, what will happen? The plants will die, the animals will die, and there we have the earth—lonely and quiet. Relight this star, and right away you re-establish the necessary cause of an infinite number of new generations, and I wouldn't venture to guarantee that with the succession of ages our plants and animals of today would be reproduced or not reproduced among these new generations.

Diderot continued to write, but the fear of imprisonment constantly plagued him. As he grew older, he rarely left his home, preferring instead to write in safety and among familiar surroundings.

When he looked out of his window onto Paris's streets, Diderot saw a world of creatures formed by adaptation. From the pigeons on the ledges to the rats in the sewers, each animal had adapted over time to fill its own little niche. As had Diderot himself. As do we all.

When he died in 1784, Diderot was buried in a crypt at the Church of Saint-Roch. The church was ransacked over the years, most notably during the French Revolution and then again a century later. The remains inside the crypt were separated and scattered.

Like a set of Legos.

What else can you build with Legos? Mountains perhaps? Continents? In the next chapter, we'll take a slight detour from the creation of plants and animals and consider something else that has been altered by gradual changes over time. That something else is the very ground we stand upon.

Chapter Seven

A Slow-Moving Ocean

James Hutton began his career as a physician's assistant. He attended University at Edinburgh and earned his doctorate in medicine at Leiden in 1749. You would think this would have led him to pursue a career as a physician. Instead, James Hutton studied nature. Particularly rocks. Not the most exciting objects to study, but Hutton saw something special in them. He saw the earth's great hidden history that spanned far more than a few thousand years. It was millions of years old. Aristotle would have nodded in approval and told Hutton he was on to something. There is something special about the ground we stand upon.

Hutton left medicine behind and immersed himself in the study of the earth and its strata. He theorized that the earth's crust, and its features, evolve and change over time. Vast amounts of time. Too vast to contemplate. The land and its deep layers move, are thrust upward, and settle down, only to be covered over by more upheavals and changes. The land we stand upon is like an ocean except that it moves so slowly as to be virtually imperceptible. If one were to look close enough, the evidence is there, hidden in the rocks. Recorded within those rocks is a history waiting to be discovered.

Hutton found inspiration in the slow changes the earth had undergone. Here was a planet composed of rocks and minerals which had endured catastrophic changes, decay, and destruction, only to repair itself. The life upon it witnessed and suffered through these changes as well. Change is inevitable. He saw this in the rocks and in the land he farmed.

It took Hutton twenty-five years to reveal his thoughts in a work titled *Theory of the Earth; or an Investigation of the Laws observable in the Composition, Dissolution, and Restoration of Land upon the Globe.* It established his reputation as a thinker and a geologist. This was in 1785, and, for the last twelve years of his life, he would dedicate himself to building upon the ideas presented in his work.

Buried in Hutton's 1788 edition of *Theory of Earth* was a section called *Theory of Rain.* He talked about rain—and something else.

If an organised body is not in the situation and circumstances best adapted to its sustenance and propagation, then, in conceiving an indefinite variety among the individuals of that species, we must be assured, that, on the one hand, those which depart most from the best adapted constitution, will be the most liable to perish.

It wasn't only the earth beneath our feet that changed. The creatures upon it, those that had to live through the earth's upheavals and changes, did as well. Hutton didn't stop there either. In a 1794 work titled *Investigations of the Principles of Knowledge*, he pushed the idea a little further:

While, on the other hand, those organised bodies, which most approach to the best constitution for the present circumstances, will be best adapted to continue, in preserving themselves and multiplying the individuals of their race.

Adaptations that provide benefits are passed on to an organism's offspring.

All this from a geologist.

What Hutton had done was to provide a theory for the changes we see around us. Not only as it affected the contours of the land, but also in the shapes of the animals that moved and fed and struggled upon that land. What he didn't propose was a theory for the creation of *new* species. That was for the Designer's hands only. Again, Aristotle would have been pleased. Changes in the land couldn't possibly give rise to an entirely new creature; it could only account for the diversity witnessed *within* a species. There are many varieties of squirrels, and that speaks to the power of the mechanism he identified. What Hutton didn't see was how it could, over vast periods, mold a squirrel into something completely different. The Designer might have developed this mechanism for change to provide a species with a means to adapt to its surroundings, but the creation of a new species was reserved for Himself. Mountains formed, landmasses separated, volcanoes melted and molded the land, and animals needed the ability to adapt to these changes. The Creator allowed for that, and Hutton pointed to the evidence. It was yet another example of the truth that Aristotle had said we would find in the world around us. All we had to do was look, overturn a few stones, and it was all there.

The Designer, the one who started it all, is an artist who has hidden the answers to all questions within His wondrous works.

If land was a slow-moving ocean, so were the things that emerged from it. They spread out, struggled to survive, and adapted to overcome the obstacles they faced. Their limbs might elongate or shorten. Their eyes may adapt to see both above and below the ground. Their skin's pigment may change depending on their environment. Still, they didn't transform into something new.

While these changes may create varieties that looked different from their neighbors on the other side of an impassable mountain, these changes were only skin deep. Hutton was not alone in his assertion that the environment a species found itself in was responsible for the changes it underwent. There was another man who cast a curious eye upon the differences he saw in the animals that jumped from branch to branch or scurried about on the ground.

This man was American-born W. C. Wells. While Hutton was busy composing his theories about the earth, Wells packed his things to move to England. It would be an eventful move, although, unfortunately, one that would almost go unnoticed by history.

Chapter Eight

It's Only Skin Deep

William Charles Wells was born in South Carolina in 1757 of Scottish parents. He might have been a product of the New World, but his soul was inexplicably drawn to the Old World, and it refused to let him go. Throughout his life, he would travel back and forth between the two continents, chasing and changing professions, as well as his fortunes.

The first time he left North America to return to his ancestors' land was to study in Edinburgh. This led to a career in medicine, which Wells pursued in cross-Atlantic trips for several years. He served as a medical apprentice in Charleston, South Carolina, and studied to be a surgeon in London. Like many before him, and since, Wells's initial field of study didn't suit him. Medicine was fascinating, but that fascination was short-lived. Something was missing; he just wasn't quite sure what it was. To find it, he packed his bags, went back to America, and settled down in St. Augustine, Florida, to become, of all things, a publisher. He was so taken by the art of publishing that he established the very first weekly newspaper to find its way into Florida households called the *East Florida Gazette.*

It still wasn't enough. Never one to be tied to a single place or profession for long, Wells retired the *Gazette* after one year and moved back to England to resume his medical practice. This time, he told himself he was going to stick it out. It was time to settle into something permanent.

It is here that our concern with the life of W. C. Wells gets interesting.

As his reputation in medicine cemented itself, Wells turned his attention to writing. His time as a publisher provided him with a unique view of the written word. He began to produce scholarly papers, but the one that gained Wells recognition and earned him membership into the Royal Society of Edinburgh was an "Essay on Dew." It's exactly what it sounds like—an essay about the morning moisture you find on grass. Not exactly the most exciting of subjects, at least as far as the general public is concerned, but it made an impression with academics. Back then, before Wells presented his paper in 1814, people thought the cold air one experienced in the morning was caused by the dew and frost on the ground. Curiously, nobody asked how it got there in the first place. Wells tested and concluded it was actually the temperature of the air and the ground that caused the dew and frost to form and not the other way around. The 1800s were quite heady times for science.

When Wells published his paper, he included another essay presented to the Royal Society the previous year. This one had nothing to do with dew but had everything to do with one's skin color. The connection between the two was anyone's guess. In it, Wells proposed that the color of someone's skin does not, nor has it ever, indicated a great divide between races as was popularly believed.

This paper's origin came about after observations Wells made upon tending to a young patient with a skin problem. The more he thought about it, the more he realized that differences between people with dissimilar skin colors were no more than skin deep. Even the most well-trained anatomist would not see any difference if one were to look beneath the skin. Wells liked to compare it to someone with resistance to smallpox. If you were to stand two people in front of an expert on the disease and ask them to identify which was resistant and which one wasn't, the expert would fail to do so. There is nothing externally observable when it comes to that resistance.

Wells grabbed hold of this observation and ran with it. That paper, the one he slipped in with his scholarly paper about dew, was titled, "An Account of a White Female, Part of Whose Skin Resembles That of a Negro." It included the following passage.

What was done for animals artificially seems to be done with equal efficiency, though more slowly, by nature, in the formation of varieties of mankind, fitted for the country which they inhabit.

Here, in no uncertain terms, Wells pointed out that mankind is not immune to nature's ability to modify an organism's features over time. Remember Hutton's assertion that slow changes in the land affected the creatures that lived upon it? Wells didn't stop there. In his paper, he went on to say:

Of the accidental varieties of man, which would occur among the first scattered inhabitants, some one would be better fitted than the others to bear the diseases of the country. This race would multiply while the others would decrease, and as the darkest would be the best fitted for the [African] climate, at length [they would] become the most prevalent, if not the only race.

While the theory Wells presented didn't account for new species, it certainly shed light on the diversity one finds *within* a species. Wells took Hutton's observations a step further and applied these changes to man. Animals weren't the only ones affected by the environments and climates that separated members of the same species. Mankind too, he explained, was as well.

Sadly, this was to be the only treatment Wells was to ever write on the subject. Three years after having been elected to the Royal Society and receiving the esteemed Rumford Medal for his dew essay, W. C. Wells died of heart failure on September 18, 1817. Any further thoughts he may have had on the subject died with him, and his paper on skin pigmentation went unnoticed and was soon forgotten.

Despite this setback, things were gearing up for some significant changes in our understanding of the vast tapestry of life. It's time to meet a man whose last name will be familiar but whose first name you may never have heard.

Let me introduce you to Erasmus Darwin.

Chapter Nine

From the Sea

Bones fascinated Erasmus Darwin. We're not talking about normal, everyday bones. We are talking about old bones. Very old bones. Fossils, in fact. Bones that pop up every so often when the earth is turned over. In the second half of the eighteenth century, every fossil found was a cause for celebration and amazement. What were these strange and marvelous creatures that nobody had ever seen before? These were the exact questions later to be raised when Mary Anning began to pull the fossilized remains of bizarre and unknown creatures out of the cliffs at Lyme Regis.

As far as Erasmus was concerned, he wasn't content to admire them in a box; he wanted to see them in their natural state. In the ground. Although he wasn't an archeologist (there wasn't even a term for one back then), when word reached him that a fossil was found, he would travel from his home in Lichfield, England, to see it before it was pulled out of the earth and put on display. Fossils spoke to him as nothing else ever had.

Life as a country physician enabled him, and others such as James Hutton and W. C. Wells, to venture into unexplored areas of science under the guise of medicine. While his duties occupied his present, fossils prompted Erasmus to contemplate the very distant past, as had rocks and minerals for Hutton. In this past creatures left bones like the ones found in the earth. Exotic creatures never seen before. He would close his eyes and imagine what they looked like when covered with flesh. He visualized how they moved and what they might have sounded like. What did they eat, and more importantly, what happened to them to cause their disappearance? A good many of these creatures were nonexistent. Like trilobites or the fossil of an aquatic monster. It baffled the senses. Some of these fossilized remains looked an awful lot like animals that were still alive. The thing was, they weren't quite the same. They were different enough to be another species or even an earlier form of a familiar species.

Erasmus wrote his thoughts down and, when he felt up to it, shared them with friends. Not all of his friends were thrilled with the direction his studies were taking him. They felt they led him to strange places. One friend, the vicar and poet Richard Gifford, told him with certainty that inquiries into matters such as those that Erasmus delved into were not pious and should be avoided at all costs. God would not be happy.

As a result, Erasmus retreated a bit and preferred to let his views on the subject be known in more subtle ways. He had scallop shells added to the Darwin family crest and included the Latin phrase *E conchis omnia*. That phrase literally means "Everything from shells." He proudly displayed this crest on his carriage and on the wax stamp he used to seal his letters.

Erasmus may have thought he was being subtle, but some, like his clergyman neighbor Thomas Seward, saw right through the alterations to the family crest. Seward called him out on it in a satirical poem.

He too renounces his Creator,
And forms all sense from senseless matter;
Great wizard he! By magic spells
Can all things raise from cockle shells.

He'd been discovered. Or, at least, his not-so-secret thoughts had been. Disheartened, he painted over the crest. He was a respected country physician, and he had a practice to run. It wouldn't do for him to rock the boat. At least not yet. All that was still to come.

Erasmus had friends who shared his views, some of whom he could sit down and talk to about them. One such friend was the geologist and naturalist James Hutton, whom we met in chapter 7. The two men spent hours talking about the age of the earth and, it is of little doubt, compared notes on what that meant for life and its humble beginnings. The world was old, much older than many had thought. The arrangement of layers and the presence of fossils, many of them unknown creatures, spoke to that age. The literature was full of such thoughts and observations, especially from French authors like Benoît de Maillet and Denis Diderot. Thoughts and observations that were considered seditious.

Erasmus had a marvelous idea. Why not write about it? He may not be ready to publicly share his thoughts on the subject of the earth's age and the fossils it occasionally offered the curious seeker, but he could still capture them. He had enough information and views on the subject to fill a book. He even had a name for it should he ever dare publish it. He would call it *Zoonomia*. Or more completely, *Zoonomia; or, the Laws of Organic Life*. Even the title suggested controversy. He would go about his duties as a country doctor but would work on it sporadically, adding pages whenever he could. It would be his masterpiece one day. Or so he hoped.

Another thing Erasmus loved to read was poetry. After some time, he considered writing his own. Emboldened by his effort on *Zoonomia*, he started another work he called *The Loves of the Plants*. It was a botanical work written in poetry. An epic poem full of sexual innuendoes. In the end, he had second thoughts about the work and decided to publish it anonymously. Again, he reminded himself, he had a practice and family to protect and care for.

The Loves of the Plants was published in 1789. Erasmus was fifty-eight. The poetry itself was seasoned with hints about his views, and he used footnotes to display his more heretical notions. He assumed the footnotes themselves would go unnoticed for the most part. One such footnote read: "Perhaps all the products of nature are in their progress to greater perfection?"

Although it was worded as a question, Erasmus felt he already had the answer. This answer he would save for his beloved *Zoonomia*. Until that time, he had a second volume of poetry to publish, a companion piece to the first titled *The Economy of Vegetation*. This was published in

1792, also anonymously, to critical acclaim. A clever man he was. With his medical practice doing well, and the publication of two successful poetry volumes, he had no trouble paying the bills. On top of this, as he grew older, he became bolder when expressing his thoughts. Riding high on his books' success, Erasmus believed he could stomach any critical backlash that might come from releasing his *Zoonomia* to the world.

Erasmus unleashed it to the reading public in 1794. It contained the sentence that was to later come back to haunt him and anyone who read it:

Would it be too bold to imagine, that in the great length of time, since the earth began to exist, perhaps millions of years . . . that all warm-blooded animals have arisen from one living filament?

To prepare himself for the outcry sure to come from *Zoonomia's* publication, he retreated to his home and waited. He grew impatient and eagerly devoured the papers searching for the harsh words he felt sure would appear.

He found nothing.

The backlash he expected and somewhat feared didn't happen. Either nobody cared, or his little seeds were so small as to go unnoticed. Perhaps they were subliminally digested by his readers and were waiting for a more fertile time to bloom?

And then it happened. The reverberations took some time, but they reached him a year later.

In 1795, a reviewer condemned the book. That was the first punch. The next flurry of punches would not be delivered until 1798. When they came, they came fast and furious. A young author named Thomas Brown published a response to the *Zoonomia* called *Observation on the Zoonomia of Erasmus Darwin*. In it, he skewered Erasmus's views.

To make matters worse, Erasmus's publisher, Joseph Johnson, was imprisoned for publishing other seditious materials. One example brought to the trial was Johnson's own periodical, the *Analytical Review*, which offered its readers radical thoughts and ideas on politics and religion. The storm Erasmus had anticipated with nervous fear and excitement finally arrived, and it was time to do another rain dance. Why not conjure more clouds to wash society clean of ignorance? He wasn't getting any younger. What could they do to him now?

His next salvo came in the form of another epic poem, one more obvious in its intentions than his previous work. It was titled *The Temple of Nature, or the Origin of Society, a Poem with Philosophical Notes.* In it, he traces the beginning of life by spontaneous generation:

Hence without parent by spontaneous birth
Rise the first specks of animated earth;
From Nature's womb the plant or insect swims,
And buds or breathes, with microscopic limbs.

And after that generation, this newly generated life adapts and changes over time.

These, as successive generations bloom,
New powers acquire, and larger limbs assume;
Whence countless groups of vegetation spring,
And breathing realms of fin, and feet, and wing.

Life eventually pops its head out of the waves to view dry land. Its next home.

As in dry air the sea-born stranger roves,
Each muscle quickens, and each sense improves;
Cold gills aquatic form respiring lungs,
And sounds aerial flow from slimy tongues.

In the accompanying notes at the end of the volume, his materialistic ideas are outlined in clear and precise prose. In the *Zoonomia*, he had asked if it would be bold to imagine that all of life had come from a single filament. In *The Temple of Nature* that tentative question is gone. He *is* bold, and he has the answer: "It must be therefore concluded, that animal life began beneath the sea."

Unfortunately, Erasmus was never to see his last work reach the public's eyes. Nor was he to hear or read the criticisms that followed. Erasmus Darwin's bones were buried in 1802. *The Temple of Nature* was released in 1803. Critics lambasted it as atheistic and immoral. Some reviewers refused to touch it for fear of becoming stained by its nature. He would have been exceptionally happy to hear this.

Or perhaps those who wrote ill reviews were afraid of Erasmus's intelligent prose. Might it persuade them to consider an alternative to the origin story they clung to? Especially when that origin story fought against the rising tide of reason for the next five decades.

There were others eager to add their voices to the fight.

Chapter Ten

Blind Moles and Dodos

Jean Baptiste Lamarck was born in Picardy, France in 1744. When he was sixteen years old, he wanted to be a soldier. After his father died, he bought a horse and rushed to Germany, where he joined the French army and found himself as a participant in the Pomeranian War. This war, considered the first world war, is known by different names depending upon where the various theaters of conflict took place. It is widely known as the Seven Years War or, in North America, the French and Indian War. At its core, it was a global struggle for power between Great Britain and France.

During this time, Lamarck became interested in botany. Plants enthralled him. For a man surrounded by war and struggle, plants offered something else. There was peace in the study of plants. He dove into his research so entirely that he eventually published his research in a three-volume work on the topic, which earned him membership in the French Academy of Sciences in 1779. After this honor, he earned the position of the Chair of Botany at the Jardin du Roi, the botanical garden of France, which became part of the Museum of Natural History. One of Lamarck's duties was to serve as keeper of the herbarium of the Royal Gardens.

Lamarck had an eye for subtlety. It was the small things in life that caught his attention, be it the silent struggles of plants or outside forces' effects. One of those forces was the moon, and it captured his imagination. The result of his study of the moon and its effect on the atmosphere culminated in the publication of his first paper.

Lamarck was proud of this work, but an idea brewed in the back of his mind. An idea that at first disturbed him. It involved the change of species over time. The notion that changes had occurred, and continued to occur, took hold of him, and he couldn't free himself from its lure. In May 1800, he released his initial thoughts on the subject during a lecture given to the Museum, and in 1801 he published a work titled *Systême des animaux sans vertèbres* (System of animals without vertebrae). For Lamarck, the key to the evolution of species was in the study of marine invertebrates. It had all started in the sea. Sound familiar?

At the time, the idea that species evolved was called *transformism*, and Lamarck believed animals did undergo transformations, but they did so in a vertical chain, a progression from lower forms to higher forms. Not so different from Aristotle's ideas centuries before. The similarities between species also struck Lamarck.

Lamarck didn't believe that species ever suffered extinction. Animals transformed into other animals. Those no longer with us had either transformed into another species over time or had been wiped out by the hands of man. True extinction was the by-product of man's cruelty to nature.

Life was not static. It fought, struggled, and adapted. A significant source of these transformative actions involved environmental changes. As surroundings changed, so did the animals forced to adapt to these changes. Thus was born Lamarck's theory of "Use and Disuse." If an animal were to use a part of its body more than it had in the past, that body part would adapt to this enhanced use. A famous example of this, and one often cited because it's easy to do so, is the giraffe. As giraffes ate leaves, they had to stretch their necks to reach the higher branches. This continued stretching would cause their necks to grow longer. The growth may be invisible, but it was there, according to Lamarck's theory. This growth would be passed on to the giraffe's offspring, and over generations, their necks would grow longer and longer.

This is the "Use" part of his argument. The "Disuse" involved those body parts that were not used or fell into disuse, like the dodo bird's wings.

The dodos were flightless birds whose wings were no longer needed. As a result, they were wiped out both by man and by the animals man brought to the islands. The dodos, unable to fly away, were easy victims.

When Lamarck looked at the environment, he saw many reasons for change. Not so different from the moon's effects on the atmosphere and plants. He used moles as an example. Their poor eyesight, he surmised, could be attributed to the very fact that they spent most of their lives underground. Their other senses needed to develop for them to survive. Moles are incredibly tactile. They can sense vibrations in the earth through nerve receptors on their skin, of which they have many. This enables them to avoid detection and predators who would find them to be a tasty snack.

So, how did Lamarck explain the presence of animals? Or the existence of life? He tore a page out of Erasmus Darwin's book for the answer. He attributed the arrival of life to spontaneous generation. Life appeared in its simplest, microscopic form by magic. From there, the environment forced it to transform itself from a simple organism to a complex creature. We still find simple organisms in the world, Lamarck would say, because life continuously appears. Time and time again. From there, nature takes its course, and the organism evolves to take its spot upon Aristotle's ladder of life.

So why, you might ask, do organisms cease in their development? Why not keep going? Lamarck had an answer for that too. Once the animal had reached a certain level of adaptation, it ceased to transform, like our little friend the mole. Depending on where the mole traveled to and the environmental pressures it faced, it might not need to change any further. Its nervous system would sustain the status quo. Subtle changes might occur, but the need for radical transformations would be over. Their behavior no longer required the body to change. As in the mole's case, their sense of touch didn't need to get any better or worse. There would not be any more changes for their offspring to receive. As he stated in his 1809 book *Zoological Philosophy* . . .

It is not the shape either of the body or its parts which gives rise to the habits of animals and their mode of life; but that it is, on the contrary, the habits, mode of life and all the other influences of the environment which have in time built up the shape of the body and of the parts of animals. With new shapes, new faculties have been acquired, and little by little nature has succeeded in fashioning animals as we actually see them.

For Lamarck, an animal's struggle caused change. Like the giraffe's neck. Were it not to push itself to reach the higher leaves, its neck would never stretch. If a mole were to live above ground and strive to see better, its eyesight would improve. If the dodo bird hadn't been content to spend its life foraging for food on the ground, its wings wouldn't have stopped working.

When Lamarck died in 1829, he did so in poverty. Upon his death, he too was as blind as the moles he loved to use as examples. His family lived upon the Academy's financial assistance, and his daughters sold everything he'd accumulated or owned at auction.

In the 1880s, a half-century after Lamarck's death, a German biologist by the name of August Weismann set out to disprove Lamarck's theory. Weismann felt that the reproductive cells of an animal were separate and distinct from its functional body cells. To prove this, he cut off the tails of mice. If Lamarck was right, then the fact that mice no longer had tails would give birth to mice with no tails. After many generations, this proved not to be the case. What Weismann missed was the fact that Lamarck never said an injury would be passed on. It was a misguided experiment, and one has to feel some sympathy for the poor mice involved.

When it came to plants, Lamarck wasn't the only one to find inspiration in the observable changes they underwent over generations. Two years after his death, a book was published that Lamarck may have found remarkable, were he able to get through it. Which, as we shall see, not many people were. Not all great ideas are couched in flowery prose.

Chapter Eleven

Of Poor Timber

Patrick Matthew was born in 1790 on a farm in Scotland, which he took over after his father died in 1807. As a result, he never finished his studies at Edinburgh. His *formal* studies, that is. When he wasn't managing the affairs of the estate, he took to educating himself.

Matthew became interested in trees and how to best grow them for the Royal Navy's benefit. Wood was used to build the navy's warships, and without them, the British empire would cease to advance. The navy required the best trees. As a result, it was the tallest and strongest trees that were selected and cut down to supply its demands. Matthew saw a problem with this. If you have a forest of trees and cut down the best ones, you are left with only weak trees, or those you would never want to use. But the forest must replenish itself, right? The navy's demands don't cease with the first crop of trees. They will need more, but all you have left are trees of questionable quality. Poor quality trees will produce poor quality trees. By culling the best trees from the forest, the inferior trees take over.

Matthew proposed a method to replenish the forest with trees of only the best quality. By systematically removing the trees of poor quality first, you could allow the best trees and the best timber to propagate. Over time you would have a forest of desirable trees. Perhaps even new *varieties* of desirable trees. The forest would be strong, as would the Royal Navy's ships.

Matthew published his ideas in a book titled *On Naval Timber and Arboriculture*. It saw publication in 1831 and, as some reviews were quick to point out, shouldn't have been published at all.

The *Quarterly Review* called some passages "pert nonsense." The prestigious *Edinburgh Literary Journal* started its review with the words, "This is a publication of as great promise, and as paltry performance, as ever came under our critical inspection." It gets worse. Not only were they not impressed, but they continued to attack its author. "He is abundantly obstinate and opinionative; tolerably ignorant of what he imagines he knows best; ill-educated, half learned, but affecting learning, and endued with unconquerable self-sufficiency, and an unequaled opinion of himself. Of general science, accordingly, he knows little, and less of vegetable physiology, and the anatomy of plants."

When one reads the journal's critique in its entirety, it is striking that the basis of their review rests upon Matthew being educated or articulate enough to undertake the effort of writing a book, and not of the ideas presented within.

Matthew was heartbroken. There were a few positive reviews, but they were tepid in their praise. Fewer still were those who noticed something else in his book. A certain passage that appeared in appendix B.

There is a law universal in nature, tending to render every reproductive being the best possibly suited to its condition that its kind, or that organized matter, is susceptible of, which appears intended to model the physical and mental or instinctive powers, to their highest perfection, and to continue them so.

This law sustains the lion in his strength, the hare in her swiftness, and the fox in his wiles. As Nature, in all her modifications of life, has a power of increase far beyond what is needed to supply the place of what falls by Time's decay, those individuals who possess not the requisite strength, swiftness, hardihood, or cunning, fall prematurely without reproducing—either a prey to their natural devourers, or sinking under disease, generally induced by want of nourishment, their place being occupied by the more perfect of their own kind, who are pressing on the means of subsistence.

The start of that section with the words "There is a universal law in nature" should have caught everyone's attention. Twenty-eight years before Charles Darwin would publish his masterpiece on the subject of evolution by natural selection, Matthew provides an outline of this law. It's all there. He states that nature molds reproductive beings to their highest perfection. This modeling of physical, mental, or instinctive powers provides some individuals with a survival advantage. Individuals who are unable to survive in an environment long enough to reproduce are pushed aside with "their place being occupied by the more perfect of their own kind." The attentive reader would have sat up straight when the implication of it all sank in. Unfortunately, there were no attentive readers. Matthew's two paragraphs went unobserved. For almost three decades, his book collected dust, and Matthew probably thought very little of it.

Now imagine this. It's March 1860, and it's late at night. There's a chill in the air. Matthew has started a fire and is settling down to read the new issue of the *Gardener's Chronicle and Agricultural Gazette,* which was delivered earlier that day to his door. With a glass of brandy at his elbow, he scans the paper until his eyes fall upon one article in particular. The article is titled "Darwin *On the Origin of Species.*" He reads the article and sees, reflected back at him, his own ideas, now three decades old. If he hadn't had the mind to publish them in his own forgotten work on naval timber, he would have only his word as proof. After what must have been a sleepless night, he fired off a letter to the *Chronicle* to voice his concern and hope for a little recognition.

The *Chronicle* published his letter on April 7, 1860. Another set of eyes would read Matthew's letter, perhaps on a similar chilly night, and with the same reaction that Matthew must have had a month earlier.

This set of eyes belonged to none other than Charles Darwin himself.

So what was Darwin to do? He couldn't exactly ignore Matthew's letter. It had been published, and it was there for all to see. His book had only been published six months before. A letter such as this could do considerable damage to his reputation if left unaddressed. Darwin was an honorable man, and so composed a letter in response.

I have been much interested by Mr. Patrick Matthew's communication in the Number of your Paper, dated April 7th. I freely acknowledge that Mr. Matthew has anticipated by many years the explanation which I have offered of the origin of species, under the name of natural selection. I think that no one will feel surprised that neither I, nor apparently any other naturalist, had heard of Mr. Matthew's views, considering how briefly they are given, and that they appeared in the appendix to a work on Naval Timber and Arboriculture. I can do no more than offer my apologies to Mr. Matthew for my entire ignorance of his publication. If another edition of my work is called for, I will insert a notice to the foregoing effect.

If you were the publisher of the *Chronicle*, imagine your delight in knowing that this exchange played out in the pages of your own paper. Not the *Times* of London, but the *Gardener's Chronicle*. The *Chronicle* immediately published Darwin's letter in the following issue. The publishers then awaited a response from Patrick Matthew.

They received one.

I have not the least doubt that, in publishing his late work, he believed he was the first discoverer of this law of Nature. He is however wrong in thinking that no naturalist was aware of the previous discovery. I had occasion some 15 years ago to be conversing with a naturalist, a professor of a celebrated university, and he told me he had been reading my work "Naval Timber" but that he could not bring such views before his class or uphold them publicly from fear.

True to his word, Charles Darwin acknowledged Patrick Matthew's discovery with the publication of the third edition of *On the Origin of Species* in 1861. He placed this acknowledgment in his appendix to that volume.

In 1831 Mr. Patrick Matthew published his work on "Naval Timber and Arboriculture," in which he gives precisely the same view on the origin of species as that (presently to be alluded to) propounded by Mr. Wallace and myself in the "Linnean Journal," and as that enlarged on in the present volume. Unfortunately the view was given by Mr. Matthew very briefly in scattered passages in an Appendix to a work on a different subject, so that it remained unnoticed until Mr. Matthew himself drew attention to it in the "Gardener's Chronicle," on April 7th, 1860.

So, it was out there. Patrick Matthew had arrived at natural selection being responsible for the transformation of species thirty years before *On the Origin of Species* was published. Unlike Darwin, Matthew wasn't content to give full credit to nature herself. He felt that there was another hand involved in the design of nature. A hand that pushed things along.

In 1871 he wrote, in a letter to Darwin:

A sentiment of beauty pervading Nature, with only some few exceptions, affords evidence of intellect & benevolence in the scheme of Nature. This principle of beauty is clearly from design & cannot be accounted for by natural selection. Could any fitness of things contrive a rose, a lily, or the perfume of the violet.

Fair enough. But not everyone was to agree on a Creator working behind the scenes, as we shall see in the next chapter.

Chapter Twelve

An Absent Creator

As children, the Chambers brothers devoured books. The younger brother, Robert, was especially fond of them. They were companions in a world that made little sense. Their father worked hard, and the boys had a roof over their heads and food on the table, but little else. Their only luxury was a set of *Encyclopedia Britannica* their father had brought home one day. That set was to become Robert Chambers's bible. The first volume was his Genesis story, and the last volume was his salvation. Everything in between was gospel.

When they were old enough to do so, the brothers became booksellers. This was in 1818, in Scotland, when Robert was 16, and William was 18. Robert, however, wasn't content to only sell books. He also wrote them—relentlessly.

By 1830, Robert was unhappy with how things were. He and his brother barely scraped by, and he wanted more exposure. His writing deserved a wider audience. He had something to say. The problem was, outside of the family, he had no one to say it to. In 1832, Robert and William launched *Chambers's Edinburgh Journal*. By charging only a penny, they ensured it was affordable to most anyone interested in reading it.

The journal did well, and the public bought as many copies as the brothers produced. Not everyone was thrilled with its progress. The tone of the journal didn't sit well with the church. Because it was cheap to buy, and because many read it, the brothers were asked to make it more palatable to polite society. The voice it contained was deemed subversive and atheistic.

So, what did they do? They ignored the church and anyone else who asked them to tone it down.

Robert loved science, and he loved writing about it. Most subjects, that is. One he wasn't keen on, and that he avoided, was the transformation of species. He couldn't see how it was possible. He believed there was a Creator, but not the Creator the Christian Bible espoused. His Creator created the laws and set it all in motion. There was no dipping back in to set things right or to save a few individuals from poverty or death. Robert's god created the world, gave it a good healthy spin, and settled back to watch.

To further his education, Robert gathered books on philosophy and science. He stitched everything he learned into the form of a story. A creation story. Surprisingly, even to himself, his creation story even included the transformation of species. He was warming up to the idea. Perhaps Lamarck was right. Perhaps we emerged from the sea to gradually take over the world with millions of years of modifications to stand upon. Geologists Charles Lyell and James Hutton taught that the world was billions of years old. Its geology changed slowly over time into the landmasses mankind walked upon. Perhaps the same went for animals, and they too changed over great spans of time. It could be possible, he had to admit, that new species were formed, survived, or died out. Hadn't the Lyme Regis fossils discovered by Mary Anning over the last few decades hinted as much? He had to remind himself it had been the fossils of creatures no longer with us that had inspired Erasmus Darwin. Robert could now see why.

These ideas went into the pages of a book he was writing. His description of an ever-changing universe would change the world. In 1844, he put the last word on the final page and decided it was ready. But there was a problem. He couldn't publish it under his name. He had a growing family to protect. When all was said and done, he and his wife Anne would have eleven children. He couldn't jeopardize his family's well-being to publish a book, even if the book contained his heart and soul.

The only option open to him was to do what many nervous authors had done before him, and that was to publish it anonymously. It might pain him to do so, he was immensely proud of it, but it was the only way to get it out there without causing his family hurt and embarrassment. He turned to his good friend Alexander Ireland to help him approach publishers. Ireland, a journalist, had many connections in the book publishing world and agreed to help. To further distance himself from the book, Robert asked Anne to handwrite each page before sending them to potential publishers. She agreed and, with Ireland's assistance, they sent the book off. The publisher John Churchill eventually expressed interest. Churchill mainly published books on medicine and anatomy. Still, there was something in this new book, this book about the creation of everything, that piqued his interest.

Chambers's book was published as *Vestiges of the Natural History of Creation*, and the reaction was immediate. It flew off the shelves and into the homes of readers everywhere.

These readers were enthralled. Everything was there from the birth of the universe, the formation of planets and suns, to the shaping of our little world. Things went from simpler, less complex forms to more complex forms over time. From the dust of creation rose life. The book said that life spread out and evolved into the creatures we see around us today, including ourselves.

Chambers watched all of this from the safety of the shadows. There were rumors about the author's actual identity. His name was tossed around as well, but he vehemently denied it. Many years later, he was asked why he wished to remain anonymous. He pointed to his house and said, "I have eleven reasons."

What made many readers uncomfortable were the questions he asked. How could there be something instead of nothing at all? It would be much easier for there to be nothing. Nothing would seem to be the default state of things. What natural law could there be to explain the sudden appearance of a universe? Especially when, before the appearance of that universe, there had only been an empty, endless void? It is easy to turn to a Creator as the cause. *Something* had to will the universe into existence. It *feels* right. But Chambers wasn't content to let it rest there.

Previous attempts to explain the presence of the universe and life centered around how things might have evolved. Some of those attempts even allowed for a Creator, but not necessarily. It was all cloaked in mystery. Even Lamarck's speculation on an organism's ability to develop certain traits by "use and disuse" was steeped in conjecture. What Chambers and his book brought to the table was science. He wasn't a scientist, but he consumed enough knowledge of the sciences to piece it all together in ways no one had before. At least no one with a gift for the written word.

How can we suppose that the august Being who brought all these countless worlds into form by the simple establishment of a natural principle flowing from his mind, was to interfere personally and specially on every occasion when a new shell-fish or reptile was to be ushered into existence on one of these worlds? Surely this idea is too ridiculous to be for a moment entertained.

If God was behind it all, and he was to step in once in a while to interfere, it would mean his creation was less than perfect. To say He did was to admit the world needed to be tweaked and that He had made a few mistakes during its creation.

If God were involved, it could only be as the initial lawgiver. We were simply an experiment. The universe was a watch that God proudly wound up and left to tick.

And what of the origin of species? Chambers had thoughts on this as well. He joined hands with those before him in that he felt life evolved from previous species, but not in a gradual way. In Chambers's view, it happened almost suddenly. A lion might inexplicably give birth to a creature that resembled a hyena. The hyena's offspring might also appear to differ from its parent. Changes happened in jumps and starts. There were no slight modifications. History was rife with sudden appearances. But, Chambers was keen to add, these appearances did not require the hand of a Creator. They only needed the Designer to create the laws by which the process worked.

While the public might have been thrilled about the book, the scientific community turned its back. A book by an anonymous author wasn't a book they were required to take seriously. The science might be sound in places, but in others, they would quickly point out that it fell dismally short.

The church wanted nothing to do with it and demanded edits. They also asked that the author be exposed and publicly denounced. Reverend Adam Sedgwick, the Vice Master of Trinity College in London, condemned *Vestiges* as "rank materialism." It was the forbidden fruit on the tree of knowledge, and this one contained a worm disguised in pretty prose.

In 1845, Robert put together a small response to his critics, which was also published by Churchill as a companion piece titled *Explanations: A Sequel to the Vestiges of the Natural History of Creation*. It, too, was attributed to the *Vestiges* anonymous author.

One man watched the storm from afar while closely guarding his ideas about the transformation. This man was Charles Darwin, and the storm that surrounded Chambers's book concerned him. He saw what the conservatives, the church, and the science community thought of it and what they thought of its anonymous author. Without evidence for his views, he dared not speak them aloud. *Vestiges* and its follow-up were warnings to remain silent.

There can be no doubt the book made things a little safer for Darwin in the years that followed. Chambers was one of the first to step into the waters of creation and declare them to be hot. Scalding, in fact. But the deeper he went, the more his ideas spread. It made the ideas to follow more palatable.

Darwin recognized this. In the appendix of the fourth edition of *On the Origin of Species*, Darwin acknowledged the author and *Vestiges*: "In my opinion it has done excellent service in this country in calling attention to the subject, in removing prejudice, and in thus preparing the ground for the reception of analogous views."

Robert Chambers died in 1871. After his death, his brother wrote a biography about him. He left out one pertinent fact—that Robert was the actual author of *Vestiges*. That secret was to remain a secret until William died in 1883. In 1884, Alexander Ireland, the man who had helped Robert find a publisher forty years before, finally revealed the book's true author in its twelfth edition.

The seeds had been sown. It was time to reap the harvest. Even if, as we shall see in the next chapter, that harvest involved peas.

Chapter Thirteen

The Gardener's Peas

My scientific studies have afforded me great gratification; and I am convinced that it will not be long before the whole world acknowledges the results of my work.

—Gregor Mendel

Gregor Mendel was born Johann Mendel on a farm in Austria in 1822. The Mendels had run the farm for over a century, but the family did not have much money. As a boy, Mendel found many things fascinating, and he could often be found studying bees and gardening. He was an adept beekeeper, and he spent hours trying to figure out how it all worked. The only thing that held him back was a proper education. He knew he was meant to do more than tend the family garden.

With his education in mind, Mendel found a way to finance it. He joined the St. Thomas monastery as an Augustinian friar. There he was given the name Gregor. Even if his original reason for becoming a monk wasn't quite pious, once he'd established himself, nobody could question his faith and dedication to the order. As with anything else, Mendel threw his heart, mind, and body into it. When he wasn't learning about the order, he was holed up in his room with a book.

The monastery's abbot was well aware of Mendel's choice of study and the fact that he approached gardening more scientifically than most. He felt Mendel could help the order just as much as they could help him.

If his efforts proved fruitful, and better fruits and grapes were produced, the monastery could refill its coffers. There were bills to pay.

Along with the mysteries of nature at ground level, Mendel was also transfixed by the unanswered questions above his head. He tracked clouds and weather patterns, eventually cofounding the Austrian Meteorological Society in 1865. While he was doing all this, another aspect of nature continued to capture his attention as much as it puzzled him. Heredity.

As mysterious as predicting weather patterns was, understanding how heredity worked had its own set of hurdles. At the same time Mendel tracked storms and sought to articulate his thoughts on the subject of heredity, Charles Darwin proposed that heredity involved little things called gemmules. These gemmules passed from parent to child and contained information about how to form different body parts. There was a gemmule for hands and one for feet. There were also other things to consider, such as eye color, the shape of one's nose, and the color of one's skin. A child might receive a gemmule from each parent that contained eye color. One might express itself while the other one floated around in a child's bloodstream, waiting to be passed on to offspring. That's why certain traits may skip a generation, or even generations, before showing themselves. This theory was known as pangenesis, and Darwin continued to work on it throughout his life.

While Darwin scribbled notes on the subject in England, Mendel had a little more luck in Austria.

The concept of "blending" was another theory popular among biologists at the time. It, too, assumed traits were inherited from both parents, but unlike Darwin's gemmules, these traits didn't dictate precisely what would happen. Say, for instance, your father had large feet, and your mother had small feet. According to Darwin's theory, there was a gemmule for each specific feature of a foot. Each toe may have its own gemmule. So your big toe might come as the result of your father's gemmule, your pinky toe from your mother's. What blending suggested was your features were a blend of both parents. Your big toe wasn't specifically your mother's or your father's; it was a blending of both. This blend was unique to you. You would one day pass it on to your children, who would blend traits from you with traits from their other parent. It was what made each of us unique.

Mendel wasn't buying this. It didn't fit with what he saw, so he spent more and more time studying it. He started with mice, but St. Thomas's abbot wasn't particularly keen on one of his friars obsessing over animal sex. On top of that, the abbot felt mice were dirty little creatures. So, to appease the abbot, Mendel turned to a safer avenue. Peas.

Peas were much more acceptable than mice. They had the added benefit of not requiring as much room or attention as mice. The monastery's nursery was the perfect spot for Mendel to conduct his research. Peas also had something else going for them. They had easily observable traits that expressed themselves in one of two ways. For example, the pea plants in the nursery were either tall or short, had purple or white flowers, and produced yellow or green seeds that were round or wrinkled. It wasn't a blend of traits. The seeds weren't a color somewhere in between yellow or green. They *were* yellow or green. What Mendel didn't know, but what we know today, what made peas the perfect subject for study, was that this difference depended on one gene. One gene dictated the color—the dominant gene. Nobody knew this. Genes weren't a thing.

He isn't called the Father of Genetics today for no reason. Mendel's star was about to rise. Even though, like a shooting star, you would miss it unless you looked up at the right moment. Which nobody did.

Another thing that made peas perfect for study was that the plant itself has both male and female sex organs. They do not need other plants to pollinate. They can pollinate themselves. Mendel could make sure he produced purebreds by self-pollinating the plants first before cross-pollinating them.

By focusing on one trait at a time, he could track the appearance of that trait over generations, like the shape of the round or wrinkled seeds. To make sure he had purebreds to start with, he would self-pollinate the round pea plants with themselves. He did the same with the wrinkled pea plants. Round pea plants, when self bred, produced only round peas. The same went for the wrinkled peas. After a few generations of this, he consistently saw only round or wrinkled peas. These were his base. It was soon time to cross-pollinate the round with the wrinkled.

The first generation of cross-pollinated plants had round peas. This told him, early on, that round must be the dominant trait. He then took those plants, the ones with the round peas which were the cross-pollination results, and self-pollinated them again.

And there he saw his first wrinkled peas appear in the second generation.

This told Mendel the trait was there but didn't express itself until the next generation. The more he did this, the more he saw a pattern. So, he counted the results. One by one.

In all, he counted 5,474 round peas and 1,850 wrinkled peas. This came to a 2.96 to 1 ratio. Every time he did this, he found, after several generations, a similar ratio. Sometimes it might be 2.96, or maybe 2.98, but always close to 3 to 1.

This became the golden ratio; 3 to 1 is the gold standard of genetics.

So, Mendel did what any talented scientist would do. He pulled it all together and produced a paper, which he read to a group of his peers in 1865. The paper was all about his peas and his heredity experiments. There too was the golden ratio and the math that led to it.

They listened . . . and then forgot about it.

Mendel published his paper as *Experiments on Plant Hybrids*, produced 40 copies, and waited.

Nobody stood up to validate his results.

Mendel faded into obscurity.

It wasn't total obscurity. In the silence that followed, he slowly removed himself from experimenting on pea plants. Round or wrinkled, yellow or green, it didn't matter anymore. He eventually became abbot and focused his attention on the monastery. This kept him busy, too busy to worry about his peas. In 1874, about six years into his tenure as abbot, the Austrian government revoked all monasteries' tax-exempt status. Like every other institution, they had to pay the government 10 percent of the monastery's value.

What did acting abbot Mendel do? He stood his ground and refused to pay it.

When the authorities showed up to seize some of the monastery's assets, Mendel met them at the entrance, locked the door behind him, and dared them to remove the key from his pocket.

They didn't. Mendel won that battle, but he lost another. In 1884, he died of organ failure at the age of sixty-one.

Had he lived another couple of decades, he would have seen an amazing thing happen. While he may have given up his work on heredity, the rest of the world hadn't. Scientists and enthusiasts continued to make headway into uncovering the mechanics behind it. They were rediscovering what Mendel had already found. At the turn of the century, three biologists, independent of one another, stumbled across Mendel's work. Remember those forty copies Mendel had made of his pea paper in 1866? The monastery had burned many of his papers after his death, but surviving copies continued to float around. These copies fell into the hands of these three biologists almost simultaneously, as if Mendel himself were working invisibly behind the scenes.

These three were Dutch botanist Hugo de Vries, German botanist Carl Correns, and an American named William Jasper Spillman.

All three men were onto the golden ratio. De Vries may have been the closest. While working to unlock the mysteries of the ratio, de Vries concluded that traits were inherited by little particles. These were passed from both parents to the child, and similar characteristics always had a dominant one. He came across Mendel's paper at the perfect time. When he published his paper on the subject, he never mentioned that Mendel had gotten there before he did. Luckily for Mendel, Carl Correns, the German botanist, had also rediscovered Mendel's paper and recognized Mendel's barely concealed contribution to de Vries's work. Correns called de Vries on it, and de Vries, perhaps begrudgingly, acknowledged Mendel as having been way ahead of them. "The gardener plants seedlings in prepared soil. The soil must exert a physical and chemical influence so that the seed of the plant can grow." That was from a sermon that Mendel delivered himself on Easter Sunday, 1847.

Mendel, and his peas, illuminated the way for all that followed. He'd planted a seed that only needed time to germinate. As had Mary Anning and her fossils, James Hutton's study of the earth and its changes, and even Patrick Matthew's once ignored publication on trees, all these ideas would intertwine to create a tapestry that would explain everything.

Chapter Fourteen

Anticipating Darwin

William Charles Wells, Patrick Matthew, Robert Chambers . . . they were all onto something. There was a principle at work that accounted for the diverse universe of species that shared the earth with us. The thing was, we were affected by the principle as well. Wells and Matthew, in particular, recognized this hidden force at work and how it might operate to shape one form into another over a great period of time. By the time Chambers wrote his anonymous opus, religion had continued to inform us that this "hidden force" wasn't mysterious at all. One had only to open the Bible on their nightstand to find the answers. They were all there. Curious minds had only to look, and the answers would be revealed.

Other "curious minds" looked elsewhere. As sciences, both astronomy and physics were challenging what was before believed to be gospel. The sun didn't revolve around Earth as had been accepted for centuries.

Heliocentrism, the belief that our sun was the center of the universe, was dead. Newtonian physics informed astronomy and opened the heavens for the discovery of the planet Neptune in 1846. The predictive power of science picked the lock of mystery and opened the ancient, dusty door of ignorance.

Biology was a young science. While its brethren made advances, it struggled along, dissecting formaldehyde-preserved specimens and staring in wide-eyed wonder at ancient fossils pulled from the earth. What was missing, and what religious thought claimed as its own, was the truth behind life and its endless forms.

What could it all mean? Wells and Matthew had stumbled upon the meaning but failed to make it known beyond a few paragraphs. There needed to be more. Biology needed its own Newton and Galileo. It required a catalyst. A paradigm shift.

It was about to happen.

Part Two
Darwin Steps In

Chapter Fifteen

Darwin before the *Beagle*

Charles Darwin will be forever known as the man who came up with the brilliant idea that life evolved on this planet from a common ancestor and that the driver, or the mechanism behind it all, is natural selection. Evolution is as true today as it has been for the past three billion years.

And to think, had Charles Darwin not stepped on board a little ship called HMS *Beagle* as a young man of twenty-two, and sailed around the world to collect specimens and capture his thoughts in notebooks, we may never have been gifted with this idea.

To truly appreciate this idea of his, we must dial the clock's hands a little further back. Not centuries, but decades. Back to a man we met in a previous chapter. I would like to re-introduce you to Charles Darwin's grandfather, Erasmus Darwin.

Charles never knew his grandfather. Charles was born in 1809. Erasmus Darwin, a physician and a poet who loved to spend hours thinking about nature and our place in it, died in 1802, seven years before Charles was born. Erasmus was also a writer who, in 1794, published a little book you may recall, titled *Zoonomia.*

Would it be too bold to imagine, that in the great length of time, since the earth began to exist, perhaps millions of ages before the commencement of the history of mankind, would it be too bold to imagine, that all warm-blooded animals have arisen from one living filament . . .

It's that last line that says it all: "that all warm-blooded animals have arisen from one living filament." Had he used the words "common ancestor" instead of "filament," his words would have foreshadowed those of his grandson decades later. Regardless, Erasmus revealed his thoughts with that passage. We evolved from a single organism. Just one. One small, living filament, granted with the ability to self-replicate.

There is another little phrase in the book which, if Charles Darwin read it, as he most assuredly did at one time in his young life, must have sat in his imagination to germinate. It rested and waited for the right season to bloom: "The strongest and most active animal should propagate the species, which should thence become improved."

The idea was planted, and it would take a half-century for it to burst from the fertile soil of his imagination.

Charles's father, Robert Waring Darwin, was no slouch either. Robert was a respected physician in Shrewsbury, where he and his wife had six children. Charles was the fifth. His wife, Susannah, was the daughter of Josiah Wedgwood. Wedgwood and Erasmus Darwin were great friends, and it is said they arranged the marriage between Robert and Susannah. Whatever the truth may be, the fact remains that they lived happily together until she died in 1817. Charles was only eight. The Wedgwoods, his mother's side of the family, would continue to play a significant role in Charles's life, as we shall see.

Charles was born on February 12, 1809, in Shrewsbury. He would later say he enjoyed a very happy childhood with the sole exception of his mother's death. He worshipped his father, Robert, and, after his mother's death, most of his maternal rearing came by way of his older sisters.

Charles, it must be said, wasn't an outstanding student. He was easily distracted and found much of what he was taught not worthy of his complete attention. He had, by his later admission in his autobiography, a penchant for long solitary walks. He would continue this practice of walking and thinking for the rest of his life. As an older man, when he had secured his own home, he purchased an adjacent three acres of property to serve as his private walking path. It is still there today and is called "Darwin's Sandwalk." The products of these walks would fill the pages of his books.

The notion that Charles would one day change the world of science wasn't something that ever crossed anyone's mind. Least of all his father's. Robert's desire to see his son do well in his studies was thwarted continuously by Charles himself. His performance in school was nothing to write home about. When he wasn't shooting birds, he collected bugs. He found them much more interesting than anything the headmaster had to say.

As a result, his father pulled him from school, a habit he would, unfortunately, become quite a master at, and Charles was placed into the role of medical assistant. After some time at this, Charles pulled it together and was sent to Edinburgh University. It was both his father's and his grandfather's alma mater, and, like his father, he would study medicine.

If Robert felt pleased with his son's progress, any such pleasure would soon, and inevitably, be tested. Charles had decided the occupation of a physician was not for him. One can only imagine his father's disappointment.

Once again, Robert stepped in to extract Charles from school—a nasty habit I am sure both men could have done without. At this point, Robert wasn't quite sure what to do with his son. He knew the boy was clever and an excellent shot. His skill as a clever sportsman wouldn't really do much good to anyone, so it was off to Christ's College in Cambridge for Charles to learn the life of a clergyman. Charles, who had decided following in his father's footsteps as a physician wasn't a path he wanted to tread, wasn't as dismayed as one might think about the prospect of becoming a country parson. The life of a parson would be a quiet one. When he wasn't guiding those who attended his church, he could take peaceful walks and study his newfound love of natural history. At Cambridge, this interest would grab Charles by the lapels and drag him into his future.

There is one man who needs to be mentioned here, and that is Professor John Stevens Henslow. Henslow, a botanist, was by all accounts a good man. He cultivated in the mind of young Charles the seeds of interest that had been pushed aside by less inventive pursuits up to this point and introduced him to a world he had not quite appreciated to the extent he could have. That world surrounded Charles, and he had failed to notice it.

If not for Henslow's faith in his young pupil, Charles's twenty-second year would have passed quietly for him. Instead, something remarkable happened.

On returning home from my short geological tour in North Wales, I found a letter from Henslow, informing me that Captain Fitz-Roy was willing to give up part of his own cabin to any young man who would volunteer to go with him without pay as naturalist to the Voyage of the "Beagle." I will here only say that I was instantly eager to accept the offer, but my father strongly objected, adding the words, fortunate for me, "If you can find any man of common sense who advises you to go I will give my consent." So I wrote that evening and refused the offer. On the next morning I went to Maer to be ready for September 1st, and, whilst out shooting, my uncle (Josiah Wedgwood) sent for me, offering to drive me over to Shrewsbury and talk with my father, as my uncle thought it would be wise in me to accept the offer. My father always maintained that he was one of the most sensible men in the world, and he at once consented in the kindest manner.

Thus, Charles's life would take a most unexpected turn for both him and the history of science.

Chapter Sixteen

HMS *Beagle*

In 1831, Charles found himself at the start of the path that would lead him into his future. Or in this case, the foot of a gangplank. At twenty-two, Charles had never sailed before, so he didn't know quite what to expect of either the voyage or the ship's captain.

The captain was Robert FitzRoy, age twenty-six, and this was to be FitzRoy's second voyage on HMS *Beagle*.

As far as first impressions go, FitzRoy wasn't initially overwhelmed with the young naturalist who'd boarded his ship.

FitzRoy was a phrenologist, and phrenologists felt that examining certain features could ascertain a man's character and disposition. These included facial measurements and the overall shape of a person's head, bumps and all. FitzRoy, it seems, was a nose man, and he wasn't thrilled with Charles's. Charles had to convince the young captain that his nose wouldn't impede his duties or abilities to serve as a companion to the captain, and FitzRoy eventually acquiesced.

With his notebooks, gear, rifles, and trunks loaded, Charles stood on the *Beagle* deck to bid England farewell. The date was December 27, 1831. Shortly after setting sail, young Charles impressed the captain even further by falling ill with what was to become chronic seasickness. As a result, Charles spent as much time as possible onshore whenever they stopped to take on supplies or familiarize themselves with the locals. He was in no rush to reacquaint himself with his sickness.

At the various stops in South America, Charles traveled by horseback. He marveled at the tropical forests and wildlife he encountered and had a trained eye for specimens such as bugs, birds, and shells. He took as many opportunities as possible to collect them. He kept some with him on the ship to study, and others were sent back to England to wait for him when he returned. One fossil he found, packed, and sent back home was the jawbone of a Mylodon, an extinct species of giant sloth. As one can imagine, this collection of fossils, fauna, birds, and bugs took up quite a bit of space. It was yet another thing for FitzRoy to become annoyed with. Lest one think the two didn't get along, they did. They had a mutual respect for one another and did the best they could during their long bouts at sea, in close quarters, and with Charles battling sickness.

In 1835 the *Beagle* reached the Galapagos Islands, off the coast of Ecuador, and spent five weeks there. The small chain is made up of sixteen islands, and Charles set foot on four of them. During a conversation with the island's governor, the governor claimed he could identify a tortoise and which island it came from simply by the shape, color, and design of its shell. It would be insights like this that fed the flames of an idea in Charles' mind. A marvelous and most dangerous idea. He noticed things he hadn't quite seen before. Mockingbirds were plentiful on the islands, and each island had its own type. There were noticeable differences as well. Some beaks were larger than others. Mockingbird diets were also different depending on the island they lived on. Charles did what any adept young naturalist would do. He collected specimens and filled his notebooks.

On October 2, 1836, the *Beagle* docked in Falmouth, England. The five-year voyage was over, and Charles and FitzRoy parted ways as friends. FitzRoy would become the second governor of New Zealand, and Charles would make history of another sort.

One thing Charles hadn't expected was that he had become somewhat of a celebrity among his fellow naturalists. It was all thanks to the specimens he had sent back home, especially the Mylodon's jawbone. In total, he had forwarded 1,529 species preserved in bottled alcohol and 3,907 dry specimens.

Now that he was home, Charles began the exhaustive and extensive study of all he had found and all he had seen. New notebooks were filled with his recollections and placed alongside those he'd crammed full of ideas while on the *Beagle*. If, at the time, you were to have pulled notebook "B" off the shelf and flipped to page 36, you would have seen a most curious sketch. You might at first have mistaken it to be a hastily sketched tree, but, upon closer examination, you would have seen that this tree was unique. There are no leaves on this tree. Only branches. The branches tell a story. This is the first appearance of Darwin's tree of life.

As Charles pored through his notes about turtles, birds, and barnacles, an idea took shape. The governor of Galapagos could tell the tortoises apart because each island's specimens had adapted to their environment. These adaptations were not passed on to the tortoises of other islands. This was because these tortoises can't swim. So while the tortoises of one island might display shells with a raised hood that allowed their host to nibble at leaves above eye level, another set of tortoises on another island might not have this same feature. That would be because the vegetation, or at least the kind they like to eat, never grows much higher than a few inches off the ground. What if these tortoises had one common ancestor whose offspring were somehow scattered among the different islands? Perhaps by ship or storm? The different environments they found themselves in would, in time, force them to change.

Charles knew this was possible. Hadn't man been doing the same thing for hundreds of years with dogs and plants? By careful selection and breeding methods, man could turn a wolf into a chihuahua and a cabbage into broccoli.

It was all well and good to say man could select for a more gentle dog with shorter hair, or corn with larger and healthier stalks. It was a different thing altogether to see how nature could do the same thing.

This troubled Charles for quite some time. Until one day, hoping to calm his mind for a spell, he sat down to read a book in his library he had beforehand neglected. One can imagine his eyes growing wider as he read.

In October 1838, that is, fifteen months after I had begun my systematic enquiry, I happened to read for amusement "Malthus on Population," and being well prepared to appreciate the struggle for existence which everywhere goes on from long-continued observation of the habits of animals and plants, it at once struck me that under these circumstances favourable variations would tend to be preserved, and unfavourable ones to be destroyed. The result of this would be the formation of new species.

In the span of one afternoon, the seed that had been planted, perhaps as far back as his grandfather's *Zoonomia*, sprouted.

The world would never be the same.

Chapter Seventeen

The Calm before the Storm

In the years following his return from his voyage on the *Beagle*, Charles settled into the quiet life of a naturalist. On all fronts, both personal and professional, things looked up for Charles. He had the pleasure of being nominated to the Royal Society of London. This distinction was only slightly overshadowed by his marriage to Emma Wedgwood just five days later. You can't ask for a much better week. The young couple didn't waste any time to start a family either. Within eleven months, their family of two had grown to a family of three with the birth of their first child William Erasmus. His middle name was a respectable nod to Charles's grandfather and author of *Zoonomia*. William was the first of nine children to grace the lives of Charles and Emma. With two exceptions, most of the children would live long lives. The longest surviving Darwin was his fifth son, Horace Darwin, who died in 1928.

Three years after the couple's marriage, Charles and Emma, with two children in tow, moved to a country house near Down in Kent. This house was referred to as the Down House, and it was here Charles would live out the rest of his life. He spent his days hovered over his notes and the specimens he had collected during his five-year voyage on the *Beagle*. He took long walks to collect his thoughts and rarely left Down House unless he attended a meeting. He preferred to correspond with his peers via letters, many of which survive to this day.

Among his close friends was the famed geologist Charles Lyell, whom I mentioned earlier as an influence on the work of Robert

Chambers. Lyell was the president of the Geological Society and wrote a book that was to have a significant impact on Charles called *The Principles of Geology.*

It was Lyell who introduced Charles to the concept of small changes over immense periods, which Lyell wrote about in his *Principles of Geology*. Although the concept was geologically focused, Charles applied it to biology.

His thoughts and notes, fueled by Lyell's ideas of change over time, developed into a small sketch of a theory. This sketch, which became a 35-page treatment, marked the first time Charles revealed the direction his thoughts had taken him. This was in 1842, and by 1844 his brief treatment had grown to 230 pages. And more was to come.

Charles was writing about transformations. Slow transformations driven by life's ability to adapt to its surroundings. If an organism's offspring were altered slightly in structure from that of their parent, and that modification was to benefit the organism's survival, those new traits would be passed on to their offspring. Accumulate enough modifications over generations and the resulting organism might bear little resemblance to its ancestor.

If Charles had hopes of making his views public, those hopes were set aside with the publication of a book in 1844 called *Vestiges of the Natural History of Creation*. The book had been published anonymously, and when Charles read it, he could see why. The book promoted the idea that life, all life, resulted from the transformation of species over time. It wasn't exactly a new idea. Mary Anning's fossils raised questions about the age of the earth and the extinct animals that once populated it, and Charles's grandfather had proposed something similar in his *Zoonomia*. The author of this book took it further by adding what he claimed to be evidence of these changes. We now know, as you recall from an earlier chapter, that the author was later revealed to be the Scottish journalist Robert Chambers.

What Chambers suggested was that everything in the universe, and he meant *everything*, developed from earlier forms. The universe begets planets; the planets beget rocks, and the rocks eventually beget life. Life popped into existence spontaneously. Chambers claimed experiments with electricity had automatically, and inexplicably, produced flies. According to Chambers, the universe's many changes focused on the production of one unique species. That species was man.

Chambers also noted the fossil record, which was extremely thin at the time, pointed to many design flaws and instances of extinction. Suppose a Supreme Being had personally created these flawed creatures. Why is it they were not walking or scurrying among his fellow Victorians? Why couldn't one go to the London Zoo to see a Wooly Mammoth? It was because, Chambers claimed, the Creator set everything in motion and sat back to watch how it all played out. Much like a child with a wind-up toy. He wound it up, set the universe down, and watched it progress with its many bumps, stops, and starts. One couldn't blame an organism's flaws on the Creator, for he never interfered. It goes without saying, but one can logically conclude that you can't give the Creator credit for the successes either. Unless you were of the mind that the point of it all was the ultimate creation of mankind. The words Chambers wrote contradicted the scriptures, which he said were based on a corrupt and childish theology.

The backlash against this book must have caused Charles to rethink making his ideas public. At least for the time being. He needed to develop them further. What he needed to find was a mechanism for the changes he felt were obvious and evidential. So for the next few years he would turn his attention to barnacles.

As would anyone who studies barnacles day in and day out, Charles became sufficiently sick of looking at them. He therefore moved on to a more interesting animal, the domestic rock pigeon. English breeders had worked with pigeons for years, and their selection process had shaped the pigeons into many different forms. If, Charles wondered, man could do this to a species of bird in only a few short years, imagine what nature could do in millions, or even billions, of years? The resulting species would be unrecognizable from the original after such a massive span of time.

Charles dove into his study of pigeons much as he had with his barnacles. Exhaustively. He consulted with breeders and bred his own. As a result, he became somewhat of an expert. If he had turned his back on his theory about evolution and decided instead to raise pigeons, he could have easily done so. Luckily for us, he didn't. One does have to wonder if his desire to fully immerse himself in his work wasn't a coping mechanism for some of life's tragedies. In 1848 his father passed away, and two years after that, in 1850, tragedy struck again in a most appalling way. His daughter Annie lost her battle with consumption, a popular term for tuberculosis. Her loss was one Charles never fully recovered from. Annie was only ten years old and was, some say, his favorite child. Whether this is true is beside the point. What can be said is it is around this time Charles suffered from chronic stomach problems. This would afflict him for many years and may be attributed to anxiety. Not only due to the personal tragedy of losing one's child but also, in no small part, to the secrets of nature that he uncovered. He knew how unpopular his developing theory would be. He had trouble with it himself.

As if pigeons weren't enough to keep him glued to his studies, he also brought plants into the fold. One thing that had troubled him with the wide variety of island plant species he saw in the Galapagos was how they spread from island to island. Perhaps they were carried by birds, but this was highly speculative and not quite realistic. What made more sense was if they floated to the different islands. The challenge there was it would require days, if not months, to do so. During their journey, the seeds would be exposed to the elements and seawater. To test this, Charles submerged seeds in saltwater for weeks to see if they could survive a possible oceanic journey. They could. It was a remarkable discovery, and this observation joined the rest in his notebooks. It is possible and entirely probable that he could have continued in this way for many more years. He could have spent the time testing his hypothesis, filling his notebooks, and breeding pigeons.

In fact, one would think he was in no rush to publish his work. In all reality, he wasn't. Not only did he want to be sure about and supply proof to support his theory, but he also knew a backlash was in store for him when he did. He couldn't take lightly what had happened when Robert Chambers published his own work. That book had made a small splash, and the resulting wave of contempt had doused its anonymous author. All because he had put forth the notion that life had evolved. Chambers had even allowed for God's help, at least in jump-starting the process. He'd even placed man at the top of the process. Charles's theory made no such concessions, and he was reluctant to let others know this.

Life, however, has a way of thwarting the plans of mice and men. Charles's views were about to find themselves on the world stage, whether he liked it or not.

Chapter Eighteen

On the Origin of Species

November 24, 1859, may not mean a lot to you today, but to Charles Darwin, and pretty much everyone alive at the time, it meant a great deal. We will begin our story, not in 1859, but a few years before that. Four to be exact. The year we are concerned with at the moment is 1855. Back when Charles was busy studying the effects of artificial selection on pigeons and the survival capabilities of seeds in seawater. He was trying to work out a method, a natural one, for the transmutation of one species into another over significant periods of time. He might have been able to affect changes in pigeons by selecting for traits he wanted to see in the birds he bred, but that was one thing. What he was driving at, or hoping to explain, was how nature did the same thing, over vast periods, without having a purpose or an end goal. He didn't doubt this was the case, but what troubled him was how it all worked. He had no knowledge of genetics at the time. Nobody did. Genetics wasn't even a word. That word wouldn't appear until 1905 when an English biologist named William Bateson would first utter it. He didn't create the word to validate Charles's theories, a common goal for every biologist at the time, but the work of another man—German scientist and friar Gregor Mendel. Charles knew a natural mechanism was there, working in the background. It wasn't magic or divine intervention. So he worked feverishly on his theory and continued to add to the pages that filled his notebooks.

He might not have discovered the engine behind heredity, but he had a theory, a strong one, for how this mysterious mechanism responded to the world around it. That response was a simple one. At least simple to some, and not so simple for the less fortunate.

Survival.

Charles knew the more offspring an organism produced, the more chances the organism's species had of surviving. Reproduction introduced modifications. No two organisms are exactly alike. Especially when it comes to offspring. They might resemble their parents, but they aren't clones. If they were exact duplicates, it would serve as a death sentence to that particular species. Why? Because with every modification or difference comes the possibility that the modification might benefit the organism's survival. If the replication process always resulted in duplicates, the chances that all organisms would suffer the same fate or deficiency would increase. The chances of survival as a species in the long term would be practically nil. In a world of white mice, a black mouse will have a better chance to avoid detection by predators at night. There is also a strong chance the black mouse will produce more offspring than its fellow mice due to the simple fact that it lives longer. If all it has to mate with are white mice, that's okay. The chances it will pass on to its offspring the genes that result in black hair are excellent, and, in time, there will be plenty of black mice running around where there were before none at all. All it took was one small modification of a gene to produce hair with a black pigment instead of no pigment at all. Charles even had a name for his theory: natural selection. Nature selected which traits would benefit its possessor. It wasn't magic, but it was the next best thing.

Charles shared his idea with his friends, most notably the geologist Charles Lyell. One day, in 1855, a paper came to Lyell's attention. Instead of passing over it, as he most likely did with many of the papers that crossed his desk, for he was an extremely busy and popular man, this paper caused him to pause. It was the title: "On the Law Which Has Regulated the Introduction of a New Species," and it was published in the September 1855 edition of *Annals and Magazine of Natural History*. The paper's author was a man unknown to Charles Lyell at the time. His name was Alfred Russel Wallace. Wallace, a young English naturalist and explorer working in Borneo, was heavily influenced by the book written by Robert Chambers. However, he didn't know the author's name at the time. Although Chambers's book was controversial, it provided no real explanation for how the whole thing worked. Wallace saw it differently. If the organisms of a species were separated, there was a chance they might evolve in different directions. Like Charles, he knew not what the mechanism for change was, but it was there, buried in the act of creation.

Lyell, who had before been skeptical of Charles's ideas, regardless of their friendship, was impressed enough with Wallace's work to suggest Charles read it, as it seemed to resemble his own work. Lyell feared Wallace's work might trump Charles's own, for the boy had published his ideas, as evidenced in the article. Charles didn't appear too concerned. He'd been working on his theory since stepping off of the *Beagle* two decades earlier. Even if they contained much of the same logic, any similarities between the two theories were coincidental.

Content that he was on to something himself, Charles continued to develop the idea that natural selection accounted for the many species and forms of life around us. He only needed definitive proof.

Three years later, everything changed. In the time since Wallace had first published his thoughts in 1855, he'd also had the pleasure to read Malthus's book *An Essay on the Principle of Population*. Wallace reached the same conclusion Charles had. He now knew what drove the changes around us. Nature did. Nature selected which organisms survived, and it all came down to modifications and traits. His work and observations confirmed the theory's validity. Wallace was familiar with Charles's work and respected his opinion. After putting all his thoughts into a paper called "On the Tendency of Varieties to Depart Indefinitely from the Original Type," he sent it off to Charles for his opinion. If Charles found the ideas contained within his paper valid, perhaps he might help Wallace publish it.

Charles read Wallace's essay on June 18, 1858. To say he practically fell out of his chair might be an understatement. Lyell was right. Wallace's ideas mirrored his own. Not only that, but some of the terms Wallace used could have served as the chapter headings of his own book.

Charles Darwin, it should be said, was an honorable man. He knew helping the young man publish his essay was the right thing to do, even if it meant his work, which predated Wallace's by decades, might be pushed aside. He showed Wallace's essay to Lyell and another trusted friend and botanist Joseph Hooker. To Lyell's credit, he didn't say, "I told you so." Lyell instead urged Charles to present Wallace's paper and one of Charles's own at the next meeting of the Linnean Society. The Linnean Society was, and still is, the world's oldest society on natural history.

Two weeks later, on July 1, 1858, Charles Darwin stood in front of the society and presented both papers in order for them both to be prepared for publication. If Charles can be said to be an honorable man, Wallace was his equal. He was extremely pleased and honored by the support he received from Charles and the presentation of his work alongside Charles's. He also understood and supported the fact that Charles had worked on the same idea for over two decades. Charles's horse had crossed the finish line only slightly ahead of his. The massive amounts of evidence that Charles had collected to support the theory of evolution by natural selection dwarfed his own.

Following the papers' presentation, Charles gathered together all his notes, sketches, and ideas to produce and publish his book on November 24, 1859. That book was called *On the Origin of Species by Means of Natural Selection, or the Preservation of Favoured Races in the Struggle for Life.*

To say it made a splash would be an understatement. It changed the world.

Charles's masterpiece was both celebrated and condemned. Chief among his detractors were those who held the belief that the world and all the creatures on it had been created in the way outlined in Genesis. God had created the world and all of its inhabitants in six days, rested on the seventh, and from then, only six thousand years had passed. Not the billions of years Charles required for his theory to hold, and certainly not in the way he said. To suggest that man descended from apes, which is not what was suggested at all, was preposterous.

The debate had found its footing in the hallowed halls of academia and on the streets. Not only in London, but the world. With one book, Charles Darwin had caught the attention of pretty much everyone. His theory was simple, powerful, and elegant. And it made perfect sense. It didn't take long to recognize its poetic beauty once you realized there was no direction or purpose behind it all.

On June 30, 1860, a heated debate took place at Oxford University which is still referred to today. It occurred at the British Association for the Advancement of Science and was presided over by Charles's former mentor, John Henslow. Samuel Wilberforce, the Bishop of Oxford, stood up against the English biologist Thomas Huxley. Wilberforce famously asked Huxley if he claimed his descent from a monkey from his mother's or his father's side. Huxley's reported reply was perfection:

I would not be ashamed to have a monkey for an ancestor, but I would be ashamed to be connected with a man who used his great gifts to obscure the truth.

It is worth mentioning that Huxley was no blind acolyte. He didn't necessarily agree with everything Charles had to say. He had his own ideas, which he applied to the framework of natural selection. The one problem he had with the theory had to do with the evidence of its validity. Huxley was an empiricist and needed to see things with his own two eyes to believe them wholeheartedly. Since natural selection took centuries to work its magic, Huxley and science were left to piece together the evidence by examining the fossil record. This was all well and good, but Huxley wanted more. He wanted to see a species split off into two branches to the point where they could no longer interbreed. He was also of the opinion that evolution might make sudden jumps in development.

Just as memorable was the presence, and reaction, of Charles's old cabin mate and captain of the *Beagle*, Robert FitzRoy. Because FitzRoy was a profoundly religious man, one can imagine his shock when he discovered or had been told his voyage with Charles marked the birth of Charles's great idea. During the proceedings, FitzRoy's dismay got the better of him. It is said he lifted his Bible over his head and implored those present to believe God rather than man.

If only there had been recording devices back then. For more than thirty years, until he died in 1895 of a heart attack, Huxley continued to defend Charles's theory.

If you visit Oxford today, you will find a stone monument marking the event outside of the Museum of Natural History.

Charles's opposition came not only from his religious peers but also from some of his professional ones. Surprising, even to himself, was the reaction of biologist Richard Owen, whom at first Darwin looked to for support. What Owen couldn't accept was the idea that species "transformed" into others. Owen wrote a review of *On the Origin of Species* in which he called the idea that man originated from apes an absurd one. As if to add salt to the wound, Owen added that it was an abuse of science. He and Charles had little to say to one another after that.

Not to be deterred, Charles continued his work. Pleased that his idea and his theory were out there and the world didn't come to an end with the publication of his book, Charles turned his attention to the very subject that vexed many who had read it. That subject was the descent of man.

When I found that many naturalists fully accepted the doctrine of the evolution of species, it seemed to me advisable to work up such notes as I possessed, and to publish a special treatise on the origin of man.

Charles's follow-up to the *On the Origin of Species* was published in 1871. It was titled, appropriately enough, *The Descent of Man*, and it laid out his notion that man's ancestors journeyed out of Africa to populate the world. The abundance of primate species on the continent, such as gorillas and chimpanzees, only made sense. One must not forget Charles worked this all out well before the discovery of fossilized remains of Neanderthals and other extinct forms of early hominids.

Never one to sit back and relax, Charles continued to release books and to correspond with admirers and peers. On Christmas Day 1881, his health caught up with him and he decided it was time to finally lay his pen down. He complained of chest pains to his wife, pains that were never to go away. On April 18, 1882, Charles Darwin suffered a severe heart attack and died the very next day. If you would like to visit him, you can pay your respects at Westminster Abbey where he is engaged in an eternal conversation with Sir Isaac Newton, who rests only a few feet away.

From this point on I will refer to Charles as "Darwin." With the publication of *On the Origin of Species*, it became how the world would know him.

And what of Alfred Russel Wallace? How did he fare? Let's see.

Chapter Nineteen

Alfred Russel Wallace

Alfred Russel Wallace was born in 1813, the seventh of nine children. His family wasn't wealthy by any stretch of the imagination. His parents did the best they could to provide for their children. As a young man, Wallace found a job as a surveyor and spent a good deal of time outside because of this. Nature captivated him as it did another young man, an entomologist by the name of Henry Bates. The two met when Wallace accepted a teaching position at the Leicester Collegiate School. They quickly became friends and Bates passed his interest in insects on to Wallace. Wallace soon found himself collecting beetles. When he wasn't looking at his compass, he was looking at the plants around him. Beetles fascinated him as nothing else had.

The stories he read about naturalists and their travels captured Wallace's imagination. One of his favorite books was Charles Darwin's *The Voyage of the Beagle,* published in 1839 as *Journal and Remarks.* By 1848, Wallace's mind was made up. He would become an adventuring naturalist as well. Both he and Henry Bates journeyed to Brazil to collect and classify insects. The plan was to collect anything they could find in the dark rain forests and send it all back to England.

Wallace spent four years in the Amazon. When it came time to return to England, he gathered up his specimens and notes and hopped on board a ship called *Helen.* Providence wasn't smiling down on poor Wallace, and the *Helen* caught fire on the journey back and forced everyone to abandon ship. Wallace's specimens, and some of his notes, the ones he'd worked so hard on to prove himself a naturalist, were lost.

Wallace spent the next eight years traveling through the Malay Archipelago, better known as the East Indies. He continued to chase down specimens and fill journal pages with his thoughts. Much like Darwin had during his great journey, Wallace noticed something in the creatures he collected. Especially the beetles. It has been said that he collected over 80,000 beetles of many different species during this time, and some of them had been unknown. While he was quick to note the differences in beetle species, he also saw similarities. There were features unique to a species, which is how he differentiated one from another, but the lines weren't so clear when he put them side by side. Different species of beetles exhibited many of the same features, only with slight, sometimes barely perceptible modifications. The reason behind these modifications came clear after he read Malthus's extremely influential book, *An Essay on the Principle of Population.*

I say influential as this same book had served as a lightning rod for Darwin's thoughts.

It occurred to me to ask the question, why do some die and some live? And the answer was clearly, on the whole the best fitted live . . . and considering the amount of individual variation that my experience as a collector had shown me to exist, then it followed that all the changes necessary for the adaptation of the species to the changing conditions would be brought about . . . . In this way every part of an animal's organization could be modified exactly as required, and in the very process of this modification the unmodified would die out, and thus the definite characters and the clear isolation of each new species would be explained.

As Patrick Matthew would later do in his 1831 book *On Naval Timber and Arboriculture*, Malthus did here in 1798. Both men had offered the explanation that some animals survived because they were "best fitted" to do so. Animals changed, as did the conditions they lived in. These changes, Malthus explained, enabled animals to adapt to their environment. Those that did not change, and were "unmodified," would die out.

Now that the method behind the transmutation of species was coming together, Wallace collated his thoughts into a treatment. He sent them to the one man whom he felt would appreciate them and who might lend considerable influence to having them published. This was in 1858, and it was because of this treatment Darwin found himself faced with a decision that concerned his own work. After twenty years of labor, he had reached the same conclusions, and here was Alfred Russel Wallace, a man literally working in the field and the jungles, with the same ideas. A year later, after presenting his findings alongside Wallace's, Darwin would publish *On the Origin of Species* and introduce the dangerous notion that species evolved, and were not created in their present forms. This evolutionary process required vast amounts of time. Billions of years, in fact.

After the publication of Darwin's book, Wallace became one of its most vocal defenders. Upon moving back to England in 1862, Wallace spread the word, through letters and guest lectures, about natural selection's beauty and how it offered the best explanation for the variety of life and the innumerable species we see around us. His friendship with Darwin continued, and many of their talks and letters revolved around their shared enthusiasm for the theory they'd both developed independently.

There were a few areas where the two men disagreed. One was sexual selection. Sexual selection involves mating preferences. These preferences select for those modifications a potential mate finds appealing or is attracted to. Perfect examples of this are peacock males who use their plumage to attract females. Darwin felt that sexual selection was just as robust a modifier as natural selection. Wallace disagreed and proposed other alternatives to those Darwin attributed to sexual selection.

Another area of disagreement was much deeper and more metaphysical. Whereas Darwin's theories did not account for a creator, Wallace thought differently. One would require a creator, Wallace surmised, when it came to explaining human consciousness. He could not see how evolution or natural selection could account for it. Material causes were inadequate and could not be responsible. Darwin held fast and was adamant that purpose had no place in evolution. The fact that we, as human beings, have the type of consciousness we have is a matter of chance and one that might be explained at some future date. Wallace shook his head at this notion. For him, there had to be a purpose behind it. Taking it further, there might be a purpose behind everything, and mankind might be the focus or the end result of it. Whatever worked behind the scenes to see this purpose brought to fruition, whether it be God or some unseen spirit, Wallace saw evidence of it in three areas:

1. The creation of life from inorganic matter.
2. Consciousness, as expressed in the animal kingdom.
3. The further complexity of consciousness in mankind and mankind alone.

In 1864 Wallace published his account of human evolution in a paper titled *The Origin of Human Races and the Antiquity of Man Deduced from the Theory of "Natural Selection."* This came seven years before Darwin's own thoughts on the matter which he published in *The Descent of Man.* In 1869 Wallace published an account of his travels and adventures in *The Malay Archipelago*, which became extremely popular. He dedicated the book to the person who had inspired him as a young man—Charles Darwin himself. Darwin was immeasurably pleased.

Wallace is also notable for giving thought to the possibility that there may be life on other planets. In 1904 he published *Man's Place in the Universe.* In it, he concluded that the earth was probably the only place that would support life. At least as far as our solar system went. Mainly because it was the only planet that had an abundance of water. He even voiced doubt that other systems or galaxies could produce planets like our own, making ours an extraordinary place.

But there were doubts. Wallace had them in his more reflective moments, and Darwin was plagued by them. It is Darwin's doubts, which may have kept him silent had it not been for Wallace's appearance, that we'll reflect on next.

Chapter Twenty

Darwin's Doubts

Everyone has doubts. The trick with doubt is that you have to push past it. Great things can happen when you do.

May 1867 was a strange month. At least as far as the weather went in England. What started with high temperatures ended with a late spring snowfall. Charles Darwin's afternoon walks around his beloved gravel path known as the "Sandwalk" were continuously interrupted by the odd weather. As a result, he spent quite a few days inside. If nothing else, it gave him a chance to read and answer the many letters that had come in since the publication eight years before of *On the Origin of the Species*. One such letter was from his good friend and supporter, Ernst Haeckel. Haeckel was a German biologist and philosopher who was deeply influenced by Darwin's book. The two men became good friends, and it caused Haeckel to write in a letter dated May 12 of 1867:

When I see how unfairly and wrongly your great work is judged . . . how even you personally are maligned, then all my respect for the great audience of naturalists vanishes. Your extraordinary humility is seen as weakness and your admirable self-criticism is interpreted as lack of firm conviction. Of course with good, understanding and thinking men you have only gained . . . . But unfortunately these are in the minority . . . . Energetic attacks and merciless blows are everywhere necessary.

It was true. Darwin felt as if he were under attack. He received a lot of mail, and not all of it supportive. In fact, some letters he received concerned him. This was nothing new, and Darwin might have said that he was used to it. He can even be said to have expected it. It was the primary reason he'd debated with himself over the publication of his book. He needed to be confident in his views when he presented them, and he needed the evidence to be strong. It needed to withstand and repel the attacks that were sure to come.

The evidence may have been strong, but it didn't stop the aggression he faced. Whether it came from the clergy, his fellow biologists, or his readers, every adverse letter was a poisoned dart blown his way. No amount of appreciation or accolades could protect him from this.

Darwin's health took a hit. Even before he found himself thrust onto the world's stage, he dealt with extreme bouts of gastrointestinal discomfort and pain. In 1849 he temporarily moved his family to Malvern in England to try out a new water treatment cure called hydropathy. Hydropathy was developed by Dr. James Gully. Part of the treatment involved sweating and being doused with cool towels, followed by cold footbaths. The diet, too, was strict. Breakfast was biscuits and water, while dinner was often fish. There was also a lot of rest, something which Darwin apparently needed. His health gradually improved, and he felt reinvigorated. Whether it was the rest or Gully's treatment that caused his improvement is a matter of debate. No sooner was Darwin back to work than the symptoms reappeared. He went through periods of vomiting, nervousness, and debilitating headaches. He suffered through a constant feeling of fatigue and occasional dizzy spells. In 1848, he wrote to his good friend Joseph Hooker.

Indeed all this winter, I have been bad enough, with dreadful vomiting every week, & my nervous system began to be affected, so that my hands trembled & head was often swimming. I could not do anything one day out of three, & was altogether too dispirited to write to you or to do anything but what I was compelled. I thought I was rapidly going the way of all flesh.

So what was it? Unfortunately, there is no way to know for sure since we are trying to treat a patient who left this earth over 160 years ago. All we have to go on are the symptoms he recorded. Some theories speculate he contracted Chagas disease during his voyage on HMS *Beagle*. There is some credence to this since a large black bug had bitten him during the voyage, and the disease is transmitted by bug bites in the tropics. Many of the symptoms associated with Chagas are there in Darwin's notes, including headaches, fatigue, and vomiting. It can often lead to heart failure and takes many years to manifest itself. While a decent theory, it is countered by the fact that Darwin lived until the age of 73, a very old age for his time. His symptoms also lessened as he got older, not something common to a sufferer of the disease. There is also the fact that some of his symptoms, such as nervousness and fatigue, were present before he set foot upon the *Beagle*.

Life as Charles Darwin wasn't easy at times.

Perhaps a more appropriate explanation is that his troubled mind affected his health. Darwin's theory haunted him.

In an earlier chapter, we learned that Darwin, in his younger days, had been on a trajectory to become a country parson. This was at his father's suggestion when it became painfully obvious his son wasn't cut out for a profession in medicine. The chance at life as a parson had a certain appeal to Darwin. For all intents and purposes, he could lead a quiet life. After tending to his parishioners' needs, he could study God's handiwork in nature. The young Darwin was a Christian, after all. He was brought up with the knowledge that one could find God in nature if one looked long and hard enough.

That country parsonage was never to be. Instead of treading the floorboards of a church, he set his feet upon the planks of the *Beagle*. In time, the notion that God's work explained the variety of species he saw in his travels had left him. There *was* something at work here, but it might not be God. Nature had no need for divine help. Nature did just fine molding life into different and elegant forms all on its own.

These were heretical thoughts, and they troubled Darwin. He could see where his studies and research led him. They pulled him away from the church and into unchartered and even blasphemous territory.

Writing to his good friend, the botanist Joseph Hooker, Darwin said, "What a book a Devil's chaplain might write on the clumsy, wasteful, blundering, low & horridly cruel works of Nature. My God, how I long for my stomach's sake to wash my hands of it."

The book he referred to was *On the Origin of Species*. Darwin knew his ideas would lead others to question, as he had, man's origins. There was no original sin or an Adam and Eve to commit it. Mankind evolved from less complex forms. Our Adam had fins and left the sea, not a garden, over 300 million years ago.

The book's popularity exceeded expectations and fueled debate on both sides of the Atlantic. Charles Darwin, in the eyes of the faithful, had set himself up against the church.

His wife, Emma, was a devout Christian, and this by no means eased his already troubled mind. He didn't want his ideas, or the resulting storm, to hurt her. She stood by his side and, if the hurt was there, she held it well. She was supportive of her husband's ideas until the end.

I want to go back for a moment to Darwin's illness. I suspect that a form of social anxiety must have played a large role in his discomfort. Social anxiety disorders can diminish as one grows older, which seems to have been Darwin's case. Throughout his life, he suffered bouts of anxiety and often went off alone by himself to think. His work afforded him the perfect escape, and he dove into it at every opportunity. Knowing where his work led couldn't have helped his anxiety. During the time leading up to his book's publication in 1859 and the years afterward, Darwin rarely left the house. He often complained of stomach troubles and used these as an excuse to avoid travel. When Emma wasn't nearby, he became anxious or nervous.

He wrote in 1876,

Besides short visits to the houses of relations, and occasionally to the seaside or elsewhere, we have gone nowhere. During the first part of our residence we went a little into society, and received a few friends here; but my health almost always suffered from the excitement, violent shivering and vomiting attacks being thus brought on. I have therefore been compelled for many years to give up all dinner-parties; and this has been somewhat of a deprivation to me, as such parties always put me into high spirits. From the same cause I have been able to invite here very few scientific acquaintances.

Aside from his family, whom he loved dearly and completely, he loved his work. Next to Emma, it was the only thing that could calm his nerves.

Darwin wasn't the only one plagued by anxieties brought on by his work. It raised a troubling question. How could the apparent design in nature not be the handiwork of a creator? What did it all mean? As we shall see, it's a valid question, and there can be no easy answers.

Chapter Twenty-One

A Question of Design

Darwin's health problems plagued him for most of his life. Maybe they were symptoms of an illness he contracted when he was young or were caused by a form of social anxiety. We will unfortunately never know. What we do have to help us understand him better are his books and letters. From these, we are told, in his own words, that he was anxious over how his work and his ideas would be received, both by his peers and the public in general. It would be safe to say his work brought to life one particular debate that refuses to go away. It may never go away.

As a young man living in the early nineteenth century, Charles Darwin was taught in church that the universe and everything he'd ever seen, touched, or tasted, had been designed by a Creator. From the very beginning, an apple was an apple, and a horse was a horse. The horse and the apple may be imperfect representations of the perfect horse and apple (recall the chapter on Plato and Aristotle), but it harmonized with the church's view that we are imperfectly created in the image of a perfect Creator.

That Creator, of course, is God.

Nature seemed to support this, and you didn't have to go too far to see it. Watch a caterpillar retreat into its chrysalis and emerge as a butterfly, and it seems to rely on magic. Nature itself appears designed for our pleasure and our existence. The trees produce the oxygen we breathe, and the plants and streams produce the food and water we require to live. And then there is the human eye. Study it closely, and you will see what a fantastic piece of biological machinery it is. Surely that must result from a designer's handiwork. There appears to be evidence of craftsmanship everywhere we look. From human-built constructions to nature, the world seems dependent upon a design that allows it all to fit together. It breaks down at times, but even those instances of failure may be built into the design for reasons we cannot fathom. A mechanic needs something to work on, or the world would be a very boring place.

The harmony of nature, science, and God's work was clarified in the pages of a book written and published seven years before Darwin was born. The year was 1802, and the author was an English clergyman and Christian apologist named William Paley. Paley published his book three years before his death in 1805. It was called *Natural Theology; or, Evidence of the Existence and Attributes of the Deity*. It is the last part of that title that says it all. The "Deity" is God. The only God. The Christian God. In the pages of his *Natural Theology*, Paley outlined his evidence for God's existence, and it all pointed to nature. You only need to walk into your backyard to see it. Take the common honey bee, for example. A bee leaves the hive, collects nectar from flowers, and returns to the hive for the production of honey. The honey is used for food, which the bees store in intricate and perfectly designed honeycombs. When you see a slice of a hive and its honeycomb structure, there is no question in your mind that it was created and not simply a random formation. Paley pointed out the same thing goes for nature as a whole.

If you saw a stone in a field, Paley said, you wouldn't think too much of it. If asked how it got there, you might say for all you knew it had always been there. After all, it's just a stone. If you came upon a watch after seeing the rock, you wouldn't conclude that it too had always been there. You would say that someone had to have made it. It's too intricate an object to be the product of anything but design. This is precisely the conclusion Paley wants us to reach.

There cannot be design without a designer.

Young Darwin devoured Paley's words when he was in Cambridge. Paley had to be right. One would never say a watch just came to be a watch by natural processes. It was evidence of a watchmaker. So too was the case with nature. Or the human body. What an amazing construction we all are. Surely our body's intricacies are evidence of a designer? What about our eyes, for instance? They are much too complicated a design for there *not* to be a designer. Paley wasn't the first to come to this conclusion. Many others before him argued there can't be design without a designer. It would be preposterous to think otherwise. Darwin agreed with this. He knew of his own grandfather's theories about life evolving from a single filament. He had stated this in his book *Zoonomia*. There were also hints of these views in his grandfather's poetic works, which he'd disguised as botanical texts. While Erasmus Darwin's words were like tiny, planted seeds waiting to be watered, Paley's text was an entire plant delivered to his doorstep. And what a beautiful plant it was. It explained everything. To study nature would be to explore God's work and to marvel at its grand design.

Or was it?

We can imagine that once Darwin had watered those seeds planted by his grandfather, seeds preserved in a book published 15 years before Darwin was even born, his views about nature would take him into dark territories. Lands only hinted at by his grandfather.

The more Darwin learned about nature, the more he questioned everything he thought he knew. If species were immutable, meaning they never changed, how could breeders change the forms of dogs or pigeons? Sometimes the changes were subtle, while other changes were drastic. It depended on what traits the breeder selected for and how many generations the breeder had to go through to get there. What if something similar occurred in nature?

Darwin was aware of the theories proposed by Jean Baptiste Lamarck. Lamarck himself had done a 180 when it came to the immutability of species. Whereas Lamarck once opposed it, his opinion had changed at the turn of the century. Lamarck now suggested that species had evolved from prior species, and all from a single organism buried in history. It was as Darwin's own grandfather had suggested. But how? How did this happen? Was there enough time for nature to perform these wonders? New studies in geology suggested there was more than an abundance of time. The earth wasn't six thousand years old, as some of Darwin's classmates believed. It was much older. An old and aged earth would explain the many fossils that turned up in unlikely places. Fossilized shells and marine animals were found on mountains. They had either been put there by the great flood or the land itself had changed over time. Mountains may have once been underwater, and the seas had either receded or geological upheavals had pushed those chunks of land skyward. It all pointed to an old earth. Too old for the mind to wrap around. Time wasn't an issue. Those fossils had to have gotten there somehow. Unless one believed God Himself had thrust these fossils into the ground to fool mankind into thinking the earth was old. But hadn't René Descartes once said that a deceitful God would have to be an evil God? Surely God wasn't evil?

Then again—was God even part of the equation? Is it possible that nature did all this on its own?

These were the thoughts that plagued Darwin at night. They were exciting thoughts, but they were also unpopular ones. He might find a few sympathetic ears in his advisors or some of his peers, but the public at large would scoff at these ideas or sweep them quickly under the rug as they would a pile of unwanted dust.

These ideas also made him sick to his stomach, quite literally. Wouldn't it be easier to go back to his prior way of thinking? William Paley's way of thinking? According to Paley, nature required a designer. Darwin realized Nature *was* the designer. A blind designer with no goal in mind at all.

Darwin had feared the book that would excuse God from the role of Designer. He called its author a "Devil's Chaplain." That chaplain was to be himself.

Darwin laid Paley's notions to rest when he stepped on board HMS *Beagle* for his five-year journey around the world. He anxiously pushed forward, collecting specimens and recording his thoughts. The arguments he had with his companion and captain of the *Beagle*, Robert FitzRoy, would serve as practice sessions for the arguments he would endure in the future.

It was a somber day when Darwin dropped a fistful of dirt onto Paley's grave. The publication of *On the Origin of Species* saw to that. That grave needed to be much bigger now that it had to accommodate not only Paley and his ideas, but also Paley's designer.

The old argument from design in nature, as given by Paley, which formerly seemed to me so conclusive, fails, now that the law of natural selection has been discovered. There seems to be no more design in the variability of organic beings, and in the action of natural selection, than in the course which the wind blows.

If he was to survive the storm, he had to be sure. There could be no shadow of a doubt for his detractors to cloak themselves in. He wasn't alone in this. He had Emma and his children to think of. Emma supported him, but for how far and for how long? He needed to be absolutely sure of himself and his ideas. Yet he could find no other explanation.

It wasn't for lack of trying. Darwin tried to disprove his own theories and refute his own beliefs. He needed to anticipate the arguments that would confront him and the cries for evidence that would rain down upon him. Whereas his grandfather Erasmus originally cloaked his radical ideas in poetry and prose, Darwin would do no such thing. He was a man of science. This was a different time, a more rational time. When the sun rose in the morning, he pushed himself out of bed to poke and prod at nature's inner working. It was when the sun set and the doubts crept back in that he relied on the lamp of reason to light the way.

If he was looking for evidence that the Lord of the Manor was God, he was dismayed to note that all the evidence pointed to an absentee landlord.

In the next chapter, we will walk a little farther along this path. The one that led Darwin to question whether or not the Landlord ever existed.

Chapter Twenty-Two

God

Charles Darwin questioned everything when it came to the origin of species and the evolution of life on earth. His questioning led him into some pretty dark places. Areas most of us would rather not venture into for fear of what we might find. He often found exactly what he was looking for, be it in the contours of a barnacle's shell or the beak of a mockingbird. His questions often unearthed more questions—until his shovel slammed into the greatest question of all, just below the surface. He knew it was there, but he'd hoped to avoid it.

Darwin or design? That was the debate on everyone's lips in the years following the publication of Charles Darwin's *On the Origin of Species*. Darwin himself preferred to let others fight the battle. After all, he was a quiet man, and, as we've established, somewhat of a homebody. He preferred the calm solitude of his study and his books to that of a lecture hall. He'd much rather be with his family or rest while Emma played her piano. Let Alfred Russel Wallace and others speak for him and the validity of his work. Every argument in opposition to his grand theory had as its bedrock one genuine item of concern. Perhaps the only item of concern. If Darwin was right and the complexity of life evolved from one common ancestor, an ancestor who wiggled out of an ancient pond billions of years ago, then what about God? Where was His hand in all of this? Surely God's only contribution wasn't that wiggling little precursor to life. Even if God helped evolution along like a wizard behind the scenes, replacing natural selection with a bit of Divine Selection, it still didn't reconcile with the origin story that had already been established. It was there in Genesis, the first book in the Bible, for everyone to read. There, in black and white, it stated unequivocally that God created the heavens and the earth, as well as everything under the sun, including man, in six days. What of that? Where did it state in the Old Testament's opening lines that God formed the earth, sat back, and hoped for the best? The conclusion many came to was that Charles Darwin must be wrong. Horribly and heretically wrong. He must not be believed.

Wallace, while firmly in Darwin's corner, did disagree on Darwin's doubts regarding the Creator. Wallace would use beetles as an example. How perfectly they were designed for the life they led. They had hard shells to protect themselves from danger and, if need be, they could draw back that shell to sprout wings and escape. How could nature do this without help from God? Darwin reluctantly started down a path that took him deeper into nature, so deep that he saw what was really going on. Everything was connected. Everything struggled to survive, and the only way to ensure survival was to adapt. With every adaptation or modification came a test. Pass the test, and you continued to produce offspring. Fail the test, and it was off to the great unknown.

The more Darwin grew certain that nature could produce the abundance of life around us without a deity's assistance, the more he became afraid to voice an opinion on the subject. He would say how descent with modification enabled organisms to survive. He would talk about how they passed those modifications to their offspring. What he wouldn't say was what it all meant. He wouldn't publicly contradict the story of Adam and Eve. But he certainly implied it in his work. He hoped nobody would notice.

The letters came in. Some in praise and some in condemnation. Others simply curious, such as the one he received from mathematician Mary Boole. "Do you consider the holding of your Theory of Natural Selection to be inconsistent . . . with the following belief: That God is a personal and Infinitely good Being?" One can imagine Darwin sitting at his writing desk with a blank sheet of paper. He must have stared long and hard at that sheet of paper before crafting a response.

My opinion is not worth more than that of any other man who has thought on such subjects . . . . I am grieved that my views should incidentally have caused trouble to your mind but I thank you for your Judgment & honour you for it, that theology & science should each run its own course & that in the present case I am not responsible if their meeting point should still be far off.

Whether his response satisfied Mrs. Boole is unknown. In his letter, Darwin said, with considerable angst, that he found the presence of pain and suffering in the world evidence of nature's dominance over our lives. Even though suffering seemed to be in opposition to an omniscient God, he would never be so bold as to suggest that God did not exist. At least not out loud, nor in a letter. But to himself, and perhaps to his dearest Emma, he might whisper his fears that if there was a God, maybe His role was relegated to the beginning only. Perhaps he gave the earth a few good spins to keep it revolving around the sun and then left.

The idea God started it all and played no further part is Deism. Deism, or some form of it, has been around for centuries. It picked up steam in the 16th and 17th centuries with the Age of Enlightenment and the rise of scientific knowledge. Whereas mankind and his home once held a special place in the cosmos, one at the center of it all, science had shown us that perhaps we weren't as unique as we once thought. Not only did the universe *not* revolve around us, but we lived our lives in obscurity upon a mediocre planet. A ball of mud that revolved around a mediocre star that existed in the spiral arm of a mediocre galaxy. We were as far from the center as one could get and as unnoticeable. If we held any place in God's thoughts, it might be as an afterthought.

The more Darwin learned, and the more he studied, the deeper his conviction grew that humanity was simply another species of animal living within the thin biosphere of this planet. We, like every other living thing, had evolved from simpler living things.

With the publication of *On the Origin of Species*, the gauntlet had been thrown. As Copernicus, Kepler, and Galileo had done to move the earth away from the center of it all, Darwin did the same for man. If he was right, and anyone would be hard-pressed to prove him wrong, mankind wasn't that special after all. We were notable in the sense that we were the only species on the planet, perhaps the only species in the entire universe, to form the type of consciousness we enjoy. A consciousness that leads us to question our very beginnings. We are also the only species to recoil at the possible answer. From a primordial pond to a few wiggling bacterium, to an ancient fish seeking to escape predators and eat in peace, we had forged our way through the ages to the stage we are at now. Where it will all lead, no one can say. And unfortunately, there is a time limit on how far we can get. The sun won't last forever. It will eventually burn out or explode.

The future of mankind aside, this wasn't what Darwin was concerned with. His concern was of the past and our origins. He was concerned with the price his family might pay for his boldness in the eyes of man and in the eyes of God. If he had been somewhat of a recluse before, it was much easier to do so after his book's publication. He still dealt with anxiety and stomach pains. He still preferred to answer inquiries by letter and have his friends and colleagues visit him in Down rather than make the trip to London himself. He preferred to spend his days at home with Emma and the children, pursuing the next question, or adding more bricks to the citadel that was natural selection.

And of God? Darwin preferred to take the stance French mathematician Pierre-Simon Laplace did when asked by Napoleon why God did not appear in Laplace's own great book. Laplace's reply?

I had no need of that hypothesis.

The hypothesis held by Darwin evolved as well. Whereas he would have aligned himself with Deism in his younger days, by 1879, his views had changed. In a letter written to John Fordyce, author of the 1883 book *Aspects of Scepticism*, Darwin wrote:

My judgment often fluctuates . . . . Whether a man deserves to be called a theist depends on the definition of the term . . . . In my most extreme fluctuations I have never been an atheist in the sense of denying the existence of a God.—I think that generally (and more and more so as I grow older), but not always,—that an agnostic would be the most correct description of my state of mind.

Darwin's friend and supporter Thomas Huxley coined the term agnostic in 1869. It basically states that questions about the existence of God are unanswerable. An agnostic neither believes nor disbelieves.

It was the view that Darwin would hold for the last three years of his life.

As mentioned, Darwin left the arguments to others. It was up to them to explain natural selection and how it worked. For some, it was not an easy task. Life is complicated, right? Any attempt at an explanation would have to be extraordinarily complex and challenging to understand. It would take a library of books to scratch the surface. *On the Origin of Species*, however ambitious it might have been, could not possibly hope to offer a simple explanation. Right?

Wrong.

Natural selection can be simplified, and we don't need an entire book to explain it. All we need is a chapter.

Let's give it a shot.

Chapter Twenty-Three

Natural Selection Simplified

Evolution by natural selection. What exactly does that mean? Evolution is easy. The word "evolve" is used a lot. We say ideas evolve, and our understanding of nature evolves. We also say the design of cars has evolved over the years. For something to evolve means it has changed. Preferably for the better, but not always. Take the first car, arguably the Benz in 1894, or the first commercially produced automobile, the Ford Model T in 1908. Compare them to today's automobiles, and they look a lot different. Set Carl Benz's first working prototype next to a Prius and the differences are too many to count. Models and manufacturers have come and gone. There are models most of us have never heard of, aside from some car aficionados, and there are those that have persisted. The Volkswagen Beetle is a perfect example. While it has undergone many changes over the years, the inventor of the original Volkswagen would still recognize a recent model as a descendant of his own. Present in its most recent form is the same contoured body structure and round headlights.

The automobile has evolved from those early models that chugged along country lanes to the vehicles that zip by on today's highways. We can see the progress, even in models separated by only a few years. They vary slightly in appearance and in function year after year.

So why do we see these little changes in shape and form from year to year? It's because car companies consider what sells. It's all about marketing and what consumers want. A manufacturer may tweak a headlight design or the front grill and sell a few more cars because of it. It becomes a permanent part of the design, at least for a few more years. Many of today's cars roll off the assembly line with GPS units installed. If a specific stereo system or iPhone holder pleases its customers, you will see those features propagate across the industry. That's selection. It's selection driven by the buying public. Car features and designs are dictated by "market selection." You can apply this concept to anything, from phones to blenders. Set the progressive models of any appliance next to one another, and it's all there. You will see the tests, failures, and features that have lasted.

This is what Darwin found in nature. This is what he meant by natural selection. It's not that much different from the progression of the automobile. Instead of the buying public determining the direction of change, nature is doing it. Darwin recognized why things change over time. He saw the mechanism behind it. He just didn't know the "how," or the minutia of the mechanics involved. I know I need to feed my car gas for it to run. I understand the engine causes it to move forward when I apply pressure to the gas pedal, but I wouldn't be able to tell you how it all works. Open the hood, expose its innards, and I am lost.

What did Darwin do in his book? He outlined the process of change, collected facts, and connected the dots. He drew inferences from the data and put it all together to form one overarching theory. His theory sought to explain why there isn't only one species of bird, fish, or bear, but many. It applies to every living creature on the planet, from the smallest blade of grass to the largest whale. From the earliest form of life to the most complex, he outlined a process for change—change that caused organisms to evolve and the selection process that determined which features lasted and which were discarded. Sometimes the process did away with an entire species. It would then become extinct, much like the Ford Edsel or the biplane.

So what are the facts that Darwin collected? One is that for a species to survive, it needs to have the ability to reproduce. If its offspring continues the process, the population will grow. Easy enough, right? There's not much to dispute there. Animals, fertile animals, reproduce, and their offspring do the same. If left unchecked, they will multiply and grow in numbers. Let's take rabbits, for example. Rabbits have large litters. They can have as many as eleven babies in a litter. Imagine a pair of rabbits left alone in a field with nothing to stop them or their children from reproducing? It would result in a lot of rabbits.

That's our first fact. Organisms reproduce, and this allows their population to grow.

The next fact is: resources are limited. There isn't enough food to go around when an animal continues to reproduce. Think of the rabbits in the field. As they grow in number, which they will do quickly, the availability of grass to eat will diminish as the population expands. There may be trees around with lovely green leaves to eat, but the rabbits can't reach them. They are stuck eating the grass for as long as it lasts, and for as long as it can support the growing population.

We have two facts now. Animals reproduce, and resources are limited. With limited resources and a bunch of hungry rabbits, what do you think happens? They will struggle, right? All organisms have a survival instinct. They won't be content to sit around and starve. If there is a blade of grass available, they will go for it. But so will their neighbor. Whoever gets there first, whoever is the fastest, will be the lucky recipient of lunch.

The struggle for survival is an inference. We can't quite call it a fact, but we can see it happen. If there are limited food sources available, then organisms, or animals, will fight for those food sources.

There are two more important facts we can apply to this scenario. One is that every rabbit produced will be slightly different. Reproduction does not produce duplicates. Even identical twins have differences. They may be hard to see, but they are there. It is why you can walk through a crowded room and recognize a face. We are all different. I run daily, but I would never consider myself fast. My son is. He can run circles around me. He is built differently than I am. His legs are longer. He gets that from his mother's side. Which brings me to the other important fact. Most features, like my son's legs, are inheritable. As I mentioned, he gets them from his mother's side. I'm built a lot like my father. My blue eyes are from my mother. When my brother and I enter a room, there can be no mistaking that we are brothers. We are not identical, I'm two years older than he is, but we've inherited enough of our parent's traits to appear similar.

If you were to walk through the field of rabbits, you would see variations in color, size, and speed. From this, we can come to the same conclusions Darwin did in his book. We have a lot of rabbits running around and limited resources. Those rabbits who are slower or smaller may have more trouble getting to the food. Their larger and faster neighbors will push them aside. Some are so large they can reach the low hanging leaves of trees when they sit on their hind legs. The leaves are a whole new food source, and one the smaller rabbits can't reach. In time, those small rabbits will find they cannot compete for the grass on the ground, and they certainly can't reach the low hanging leaves. Starvation will set in, which will hamper their ability to reproduce. Some will die before they can do so. The larger rabbits, on the other hand, are having a grand old time of things. They nudge the smaller ones away for the grass, and they rear up on their hind legs to get to the leaves. Their bellies are full. As a result, they live longer and reproduce more. Their offspring will inherit their parents' traits. They will inherit those genes that make them fast and those that enable them to grow larger.

What do you think we would see if we were to step away for a few months and come back? More than likely, we would see that most of the rabbits can reach the leaves now, and those smaller rabbits are gone. They were weeded out by the simple fact they could not compete for resources. This is natural selection. Nobody picked which traits provided the most benefits. Nature did. It's an endless struggle to survive and to reproduce. Suppose a variation provides a benefit, and that benefit allows its owner to survive longer and to reproduce more. In that case, that variation will spread throughout the population, just like we see with our large rabbits. But now that they are all large and fast, they are no better off than when they were all small. Now everyone can reach the leaves, and those leaves are disappearing fast. There's a higher level of leaves they can't reach.

But wait! There's one rabbit able to reach that second level. It's obvious he's been doing so for a while since his belly is a little fatter than those showing signs of starvation. He'll be around for a while.

What do you think will happen next?

Now that you can see how natural selection works on a population to produce larger rabbits let's change things up. We will leave half of those rabbits in their original environment with the large tranquil field, grass, and leaves. We'll move the other half to another environment. A dark forest. In this forest we will release some wolves to make it a bit more dangerous.

These two environments, the tranquil field and the dark forest, are far enough separated that the two groups of rabbits will never find one another again. Before you go back to check up on things, I want you to think about something. We've already seen what happened in the field. The larger rabbits could get to the leaves, and over time the population had a higher percentage of large rabbits. Given more time, they were all large. Much larger than the original group of rabbits who could only eat grass.

What about the rabbits in the forest? What do you think they would need to survive? They would no doubt need to be fast to avoid the wolves. They might also need a fur color that would allow them to blend in with their surroundings. Maybe they would need to be smaller to squeeze into safe places to reproduce, care for their young, and avoid being eaten. Now let's assume all this happened. If you were to capture a few of these forest rabbits and sit them next to their tranquil field cousins, what do you think you might see now?

Two different species of rabbits, perhaps?

Explain that to someone and you've just explained natural selection. It illustrates not only how nature creates different species of rabbits but, given a billion years to work with, it can change that rabbit into an entirely different animal.

Part Three
Evolution Tales

Chapter Twenty-Four

The Beginning

In the beginning, there was nothing.

Well, we don't know that exactly. We don't know quite what there was. Call it a "singularity" or a "quantum event in a vacuum"; all we know is there was a moment when everything started. We call it t = 0, *t* being time. Whatever occurred at the zero is out of our present reach. But t = 1 is another thing. We are pretty confident we know what happened then and what has happened since.

If you haven't guessed what this is all about, I'm referring to the beginning of everything. Long before Darwin and his great idea, long before anyone had any ideas at all, there was nothing. If you were to travel out far enough on a rocket ship and keep going, what would you find when you reached the border of the universe? Or is there a border? What if it goes on forever? What could be at the end of space but more space? And if that is the case, what of its beginning? For centuries mankind has struggled with that question. How did it all start? To further complicate matters, what was there before it started?

About 13.7 billion years ago, give or take a billion, something happened. We call it the Big Bang now, but that wasn't always the case. Before 1927 we had no name for it at all. It wasn't until a Belgian priest and scholar named Georges Lemaitre proposed his theory about how it all started, and astronomer Fred Hoyle gave the theory its name, that we call it what we do today. In the time it would take you to sneeze, matter expanded from a dense seed into an entire universe. There was a lot of heat driving that expansion, and when things cooled down, particles clumped together. These clumps affected one another via gravity. For billions of years following the Big Bang, they performed a cosmic dance, revolving around one another until time and physics shaped them into orbs.

It took almost 9.1 billion years, but there's one orb in particular we really care about. You are standing, sitting, or lying on it right now. This orb is Earth. It was born about 4.6 billion years ago; that's 46 million centuries.

Shortly after its birth, the Earth wasn't exactly a hospitable place. It was hot, volcanic, and oxygen was a rare commodity. There was nothing we could call life. There were a whole lot of gases, water, and storms.

So the question now is: How did life emerge from these conditions? Unfortunately, we don't exactly know. But there are theories, and Charles Darwin tentatively offered one of them.

But if (& oh what a big if) we could conceive in some warm little pond with all sorts of ammonia & phosphoric salts,—light, heat, electricity etc., present, that a protein compound was chemically formed, ready to undergo still more complex changes, at the present day such matter would be instantly devoured, or absorbed, which would not have been the case before living creatures were formed.

This "warm little pond" that Darwin proposed might be the answer. If so, what happened in this primordial pond 3.9 billion years ago set the stage for everything that followed. It was around this time the first self-replicating organism opened its little eyes. Well, not really. It didn't have any eyes to speak of. But it was an organism, and it probably set up shop near some long-gone volcanic vent, much like hyperthermophile bacteria do today. These little bacterium love the heat and are considered extremophiles since they can survive under dangerous conditions. Some hypothermophiles have been found to thrive next to vents deep in the ocean at temperatures close to 100 degrees Celsius (212 degrees Fahrenheit)!

In 1953 a young graduate student from the University of Chicago tested whether the early composition of the planet could have given rise to the elements necessary for life. His name was Stanley Miller, and, using two flasks, he recreated the primordial pond that might have started it all. In the flasks were hydrogen methane gases along with methane and ammonia. Then, much like a modern-day Dr. Frankenstein might do, Miller fired it up with electric sparks to represent lightning.

And then he waited.

What do you think he got? Well, no creature from an H. P. Lovecraft story crawled out of the flask, but he did get a flask full of fatty acids, amino acids, and some organic compounds. The building blocks of life were there. Now he just needed a billion years or so to see what would happen.

Miller proved that, given the right conditions, conditions that were present 4.5 billion years ago, life may not only be possible—it might be inevitable.

In a *New York Times* interview in 1996, evolutionary biologist Stephen Jay Gould wondered aloud that it might not be difficult for life to emerge under the conditions that Stanley Miller replicated. Gould was explicitly talking about bacterial life.

Lightning and gases aside, there are other ideas when it comes to how life started. Lord Kelvin, an extremely famous British physicist, cast his gaze beyond the clouds when the question came up. At a meeting of the British Association for the Advancement of Science in 1871, Kelvin himself stated that "the germs of life might have been brought to the earth by some meteorite."

Kelvin might have been on to something. It's not too farfetched to think life may have come here from somewhere out there. The notion that the seeds of life were planted on our planet billions of years ago is called panspermia. It's a lovely little theory that does have some surprising support outside of Lord Kelvin's statement. Francis Crick, the co-discoverer of DNA's molecular structure, also supported the notion of panspermia. He even took it a step further and said it could be done deliberately, meaning we could do it by seeding the universe ourselves. That would be called "directed panspermia." It might sound like science fiction, but what doesn't when you are trying to figure out how it all started?

All that aside, there is some evidence to support Kelvin's idea of life coming here from beyond our atmosphere. In 1969 a meteorite slammed into the earth near the town of Murchison, Victoria, in Australia. That meteorite is 4.5 billion years old, and it is rich in amino acids, some of which are required for the formation of proteins found here on earth. It is entirely possible that meteorites such as the one found in Murchison could have repeatedly entered our atmosphere 4.5 billion years ago. When they hit the ground and that primordial pond, their cargo would have spread. Those deposits of amino acids only needed to wait for the right bolt of lightning to make things interesting.

This is conjecture and fascinating to talk about, but without a telescope into the past, that's all it will ever be. Even if we could see what happened and witness meteors that contain amino acids splashing into our ancient oceans, all it would do is move the origin of life elsewhere.

There was a lot of activity 3.5 billion years ago—activity that caused something to happen.

With the outline of the puzzle started, let's look at some of the other pieces.

Chapter Twenty-Five

Cyanobacteria

Remember Stanley Miller? He was the young graduate student from the University of Chicago who threw together some gases and chemicals. These included methane and ammonia, both of which were plentiful on our planet in its early days, and fired the mixture all up to create organic compounds with fatty and amino acids. These compounds combined not because they wanted to but because their chemistry was right, and in doing so they formed more complex compounds that continued to absorb the surrounding organic material. Eventually, these compounds would divide. Think of those divisions as offspring. The dividing microbial organisms wouldn't be perfect duplicates. They wouldn't even be close to perfect. They would be different enough for some to fare better than others. Perhaps they were better at combining with and absorbing the organic materials they bumped into now and then.

If you haven't guessed, this is natural selection. If you take two organisms that are roughly the same but different, one organism will fare better than the other because of those differences. Not always because the organism is more robust or built better. What makes something "better" is extremely subjective. Suppose its differences enable it to survive longer to replicate itself, and those replications perpetuate its unique differences, those that were beneficial for it to survive in its environment. In that case, it can be said that nature "selected" for it to prosper. In time, its offspring will represent a higher percentage of the population. Those organisms unable to compete with their growth and ability to consume resources will either perish or move on to another environment where their differences will help them flourish. And so on and so on. (Remember the rabbits back in chapter 23.)

When I say that nature "selected" the organism as the one to prosper, I don't want it to sound like the selection process is conscious or has a purpose. It's not, and it doesn't. It's a way of saying that some organisms are better than others when it comes to surviving in an environment. Selection is a process and not an action. If I were to take a funnel and dump marbles of different sizes into it, the smaller marbles would fall through while the larger ones will get stuck. Think of the funnel as nature. The funnel is not giving any thought to the marbles that move on, nor is it giving any thought to the ones which get stuck. It's simply an environment that each marble must go through, and the smaller ones move on.

So, 3.5 billion years ago, microbial organisms combined, split, and combined some more, until microbes and single-celled algae formed. And they were small. Too small to see without having a very powerful microscope in your back pocket. One of these single-celled algae-like organisms was cyanobacteria. These little guys were so good at surviving in any environment they found themselves in that they propagated pretty quickly. By "pretty quickly" I mean over millions of years. They are so good at what they do that they are still around today. If you've ever seen clumps of blue-green algae, then you've seen them.

What they did, and what they did very well, was to find a way to process the greatest resource of all in a world of microbes struggling for resources. That resource is sunlight. If you imagine the earth in its early days as stormy, then you can imagine periods of sunlight were brief. Those early microbes took what they could get. They left other micro-organisms behind, since those other organisms couldn't do much of anything with the sunlight. It's a pity for them, but good for the cyanobacteria. Remember, the cyanobacteria were single-celled organisms. These cells contain no nucleus and thus no energy center, so they are called prokaryotes. I only mention this to differentiate them from the eukaryotes, which we will get to later. The *pro* in prokaryote means "before." It's Greek. And *karyote*, or *karyon*, means "nut." So they were the first nut.

These first nuts, or prokaryotes, or cyanobacteria, loved the sun. They used sunlight to power themselves and their replication. One by-product of this activity is oxygen.

And it is here where things really get interesting.

The process of using sunlight for energy and releasing oxygen into the atmosphere is photosynthesis. Plants are experts at photosynthesis. It can all be traced back to what these miniscule critters, the cyanobacteria, could do.

It may surprise you to know the evidence is there for scientists to pick up and examine. I'm not implying little three-billion-year-old cyanobacteria corpses are lying around. Instead, they clumped together in layers, and these layers mineralized into structures called stromatolites. This happens today as well. The fact that we have stromatolites which date back 3.5 billion years is extraordinary evidence for the presence of early cyanobacteria.

Let's get back to the wonderful production of oxygen. What would we be without oxygen? Well, we wouldn't be what we are today. Advanced organisms such as ourselves require oxygen. Back then, there were no advanced organisms such as ourselves, but photosynthesis saw to it there was plenty of oxygen to go around where there was before none at all. All that was needed was for someone to use it. Eventually, someone to step up to the plate.

Things stayed pretty much the same for the next billion years. Cyanobacteria continued to create oxygen and release it into the atmosphere. And then one day, a day that will go down in history as perhaps the best day there ever was, at least as far as we are concerned, something happened. What it was may have seemed inconsequential at the time, but it would arguably prove to be the single most important event in earth's history, besides its initial formation, of course.

On this day, an adventurous single-celled protist called an archaeon was floating along and bumped into another single-celled bacterium that was passing by, and it captured it. There was no turning back now. What this protist and its captured bacteria found was that they worked much better together than alone. They formed the first team. A team of two who were now to become one. The next time this protist split, the captured bacteria split along with it. They had an appetite like no other, and what they craved was the oxygen the cyanobacteria produced. They could convert this oxygen into energy.

This process, where one cell captures another one, and the two become dependent upon one another, is called endosymbiosis. The process of splitting is called mitosis. They were no longer prokaryotes. They were something else. Something new. They were eukaryotes.

What I've just described to you is one model proposed to explain the beginning of complex cells. It's called the Chimeric model. It might also explain the presence of a cell's mitochondria, otherwise known as the cell's power plant. It's possible the bacteria that were captured became the mitochondria that power our cells today. Whatever the case may be, the eukaryotes multiplied and spread, with each new cell essentially a clone of its parent. These new cells might be complex, more so than the simple single-celled organisms they once were, but the lack of variety or changes when splitting kept them stagnant. This uneventful consumption of oxygen and cell division continued for another billion years. And it might have continued until today. The first eukaryote appeared on the scene around 2.1 billion years ago. Half of the earth's current history had passed. I say "might" have continued because something else happened. For natural selection to work its magic, it needs something to work with. Some type of change. If these eukaryotes, which were very good at what they did, were simply splitting off into clones, there was really nothing for nature to do. It could only sit around, twiddling its thumbs, and watch these organisms propagate and spread in the same dull manner they had been for a billion years.

Nature required something else to kick things up a notch.

That something else is sex.

Chapter Twenty-Six

Reproduction

Imagine this. In a sea swimming with clones, one clone drifts toward another clone. They usually just bump into one another, tip their hats, and go on their way. It's the way of the world. Their lives are pretty dull. They drift, they eat, and they split. But one day—a day, we should remember, when no one was around with a notepad to capture it—something happened. There are countless stories of love found and lost, but nothing compares to this one. At least Romeo and Juliet had a template to go by. They knew what love was; they had seen it before. They just didn't realize how powerful it was until it swept them off their feet and swept them away.

But the event that occurred to our two eukaryotes came as a total surprise. When they bumped into each other on a dark and stormy night in the middle of the sea, something magical occurred. Neither one knew it at the time, but something passed between them, and when they parted, it was too late. The deed was done. They both left that encounter slightly different. What passed, from one to the other, were a few microscopic bits of genetic material. And this encounter changed the world, because the next time they split they differed from the surrounding clones. Because of that chance encounter and the swapping of genetic material, the new cells contained a little bit of both of the cells that had bumped into each other. We'll call them their parents, for that is what they were, although they had no idea what they had done at the time. These new cells found exchanging genetic material like their parents had wasn't a bad thing to do. They couldn't control themselves when they bumped into one another. They did as their parents did and passed genetic material back and forth, creating new variations of cells that hadn't existed before.

We don't know how or why all this happened—it was over a billion years ago. All we know is that it happened, and because it happened, it gave nature a new toy to play with. I'm romanticizing and embellishing this, but no amount of embellishment can truly capture how amazing this event was. These little eukaryotic cells did not understand what was happening. They had no "ideas" at all. Perhaps their cell membranes were a bit too permeable, or maybe one clone suffered a mutation that allowed it to puncture the cell wall of its neighbor. Whatever the reason, once it happened, there was no going back.

When they were all clones, things were pretty stagnant. They competed for the same resources. Nobody had an advantage over their neighbor. If a cell's copy suffered from a mistake that happened trillions of times, the cell would quickly die out. The world didn't suffer mutations. Not when there were perfectly healthy clones around, and plenty of them.

When cells swapped genetic material all previous bets were off. Nature, who had previously slumbered out of sheer boredom, was forced to sit up and take notice. Here was something new. Among this sea of clones, new cells were appearing. These cells had some of their "mother's" genetic material, as well as some of their "father's." I'm using the terms *mother* and *father* loosely, but you know what I mean. These new cells had something from both parents. They were different. They didn't fit in with the surrounding clones. It's because they swapped genetic material that there were variations. These variations went on to swap genetic materials with their neighbors, split, and create further variations.

In some cases, the new cells failed to survive, but others had an advantage. Perhaps they found they could process oxygen more efficiently and divide more quickly and more often. While doing so, they continued to swap and mix genetic material with their neighbors. Different variations emerged, and the cycle continued.

This gave natural selection something to work with. Remember, natural selection occurs when a variation is advantageous to individuals or groups that possess the variation. Suppose they survive long enough to reproduce or replicate. In that case, that variation is passed on to some of their offspring, or all of their offspring, depending on the variation and the genes involved. With these new cells and new survival techniques, they created a population of their own. The clones weren't alone. A new species of cell had arrived, one which wasn't content to go off all by itself to split into duplicates. These new cells joined for a moment, exchanged genetic material, and divided by a process called meiosis.

Remember, before the exchange of genetic material, it was all mitosis and the creation of clone cells. Splitting off into clones is fast, and it worked very well. They did this very efficiently for well over a billion years. There is a problem with this. It's not like the early earth was a safe place. The environment was an extreme place and prone to many sudden and adverse changes. These little eukaryotic cells had this to deal with. Some pathogens drifted into their clusters to wreak havoc. As they were all clones, they all suffered the same fate. Imagine I have an allergic reaction to blue light and that, whenever I encounter it, I break out in a horrible rash and find myself confined to bed for months. And then an evil scientist, I'm not sure why he's evil, but it just feels appropriate, this evil scientist creates a clone of me. This clone is you. He then hustles both of us into a little room and turns on a blue light. We both break out in a horrible rash and are both confined to bed. What affects me affects you. There is no getting around it because you are my clone and we are exactly the same, right down to our genes.

Now let's say that instead of you existing as a clone, you're slightly different. Perhaps all of our cells, the ones that make us who we are, are *almost* identical, but when we look closely, deep within your genetic makeup, there is a variation in the code, a mutation. We appear to be the same for all intents and purposes, yet the gene that dictates our reaction to blue light is different. It doesn't affect you the same way it affects me. In fact, it invigorates you. The next time the scientist flips on the blue light, I break out in a rash and fall to the ground, yet you feel a burst of energy and run out of the room. You have an advantage here, don't you?

Let's say that the evil scientist's blue light is substituted by a blue sun. Me and all of my clones will suffer disabling rashes for most of our lives. You, on the other hand, will happily go about your way. When you reproduce you will pass your genes on to your offspring. These are the genes that allow your offspring to walk around in the blue light. Given time, the world will contain more of your offspring than it will of mine since I've spent most of my life in bed and in pain. As would all of my clones.

None of this ever happened, but similar things did. The eukaryote clones faced their own versions of the blue light dilemma, and it affected them all in the same way. Let's say the blue light gene was a real thing, and the earth's atmosphere caused the sun to appear to be blue and only allowed light from the blue spectrum to enter. You can imagine a mass extinction, except for those cells that had exchanged genetic material and, in doing so, introduced variations that might be able to cope with the blue light. That ability to cope might be a mutation that was passed on and duplicated. The point is that nature now had a way to shuffle genes in a million different ways and billions of years to do it.

Think of it this way. Take ten decks of cards, pair them into twos and shuffle them together into five piles. Flip over the first card. If it's an even card, you keep it. If it's an odd card, you toss it in the trash. Let's say two of the piles had an odd-numbered card and are thrown away. You are left with three piles. They survived. If all the piles were exactly the same, as in the clone situation, and all of their top cards were odd, they would have all been discarded.

That's what natural selection does. It's a never-ending game of cards that are shuffled and reshuffled over and over. The rules change and, when it does, piles are discarded, and some continue to be shuffled.

Up to this point, we've been talking about single-celled organisms. We had our prokaryotes, which were simple cells without a nucleus. Then we had our eukaryotes. Those were the guys who captured a passing bacterium and turned it into a nucleus. They were still single-celled organisms, but they were a little more complex.

We've been talking about advantages and nature's blind habit of selecting those cells with advantages. I say "blind habit" because it is a blind selection process. Nature isn't doing it on purpose. Selecting means that, if the organism's advantage allows it to survive long enough to reproduce and pass on its advantage, its offspring will last longer. Benefits come in many shapes and sizes.

What if two eukaryote cells that sexually reproduce found that, by hanging out together, it was better than hanging out alone? Let's imagine they are floating along, and the current becomes really rough. It sweeps away all the other eukaryotes. They are separated from one another. That would make joining to reproduce extremely difficult. They are slaves to the waves. If the waves persist, they are pushed continuously apart. But let's say there happen to be two little cells that are sticky. The shuffling of genes created sticky membranes for them. The others don't have it, so they are swept away. These two lucky cells bump into each other and become stuck together as the currents push them this way and that. They do what they do best, and that is to exchange genetic material and reproduce. Their offspring have sticky cell membranes as well. Because of this they can stick to their parent cells, as do their children, and their children's children, until they are one giant mass of floating cells. This mass works together to fight the currents and to reproduce. Every once in a while, a powerful current comes along to break a section of cells off and they drift away but continue to reproduce. Now we have two groups of cells floating off in different directions. As the eons pass, pieces break off now and then, and, in time, the seas are full of these floating islands of cells. Don't forget, these groups all reproduce sexually. They pass genetic material back and forth. If a variation or mutation occurs, it is passed on. Since these floating groups of cells are separate from one another, a variation in this group won't affect the other group. The groups become different because their cells are different. The cells become so different that when one group bumps into another group, it can no longer swap genes with the first group because things have changed. Maybe the first group's cell membranes are too hard for the second group to penetrate. Perhaps the shape group one has formed doesn't allow it to connect with group two.

What we essentially have now are the world's first multicellular organisms.

If you thought this was interesting, wait until you see what happens next.

Chapter Twenty-Seven

Multicellular

There is a lot we don't know about the steps leading up to natural selection's first attempts at creating multicellular life. From our humble beginnings as single-celled organisms to our gradual development into complex cells that swapped genes, much of it is speculation. We know these little precursors to the life we see around us had to start somewhere, and by applying what we know of life today, we can deduce what may have occurred then.

We might not know how, but we know when, or roughly when, it started. We have developed strong theories about life's progression from amino acids in ponds to complex cells.

But why do we have such a hard time with this? Surely there must be evidence of this progression to multicellular bodies? It's because these organisms had soft bodies. Remember, we are talking about cells here. They don't exactly lend themselves very well to preservation in the fossil record. When their life is over, they are obliterated. There's nothing to mineralize. Thankfully, we have secondary or tertiary evidence of their presence. The stromatolites left by the cyanobacteria are a perfect example. Stromatolites form today, and we understand how. The presence of 3.5billion-year-old stromatolites tells us it happened in the beginning as well. A billion years after the earth was formed, they appeared. And then, close to 600 million years ago, according to some models, other things appeared. Multicellular organisms.

Multicellular organisms appeared on the scene anywhere from 1.2 million to 542 million years ago. Popular consensus is somewhere between 585 million to 542 million. But what's a few million years? The fact remains that by 542 million years ago, those little eukaryotes (the ones with complex cells containing a nucleus) grouped together. And it all started with a simple prokaryote.

These cells organized themselves into three-dimensional shapes. As they did, certain forms worked better than others. Furthermore, they divvied up the work. Some cells were more specialized at performing certain tasks over others. In time they found that by working together they could move around as a colony. The sole purpose of some cells was to aid in this mobility, while other cells digested the nutrients around them and passed the benefits to the other cells in the colony. It was one great big commune of cells. Resources were shared, and their size increased. This didn't happen only once; it happened many times. Multicellularity accounts for the development of algae into plants and from complex groups of cells into even more complex bodies able to move around.

So why was this important? Anytime you give natural selection a toy to play with, it does so with gusto. It's like a kid with Play-Doh who molds and shapes it into little creatures. Some look good, so they are set aside to play with some more, while some are tossed back into the can to play with later, if at all. Those Play-Doh shapes which are thrown away are gone forever. The ones that are kept eventually harden, and, voila—we have fossils. Play-Doh fossils, but you get the idea.

Natural selection had such a grand old time with multicellular organisms because they gave it something to select for. These organisms increased in size, they moved into new areas for food, and they protected themselves against the environment. With two groups of cells competing for the same resources, one group will have the advantage.

During this time, some peculiar forms emerged. One of these forms was a strange little microbial creature called the choanoflagellate. It was unique in that it consisted of nothing but a tail and a small oval-shaped head. Directly below this oval-shaped head was a sort of collar that, as the organism propelled itself through the water with its tail, captured bacteria the choanoflagellate absorbed. These developed in the seas. The land was still a dry and bare place. Many of the organisms that popped into existence hung around on the seafloor. Many of them remained immobile, and those that did were content to extract nutrients from the surrounding water.

But not all were content to sit around. Like the choanoflagellates, other multicellular organisms discovered the joys of mobility. With mobility came greater opportunities to find food. In their search, they moved into different environments and faced new challenges.

How do we know any of this? A peculiar thing happened a little over 500 million years ago. During the Cambrian period, an explosion of animal life was recorded in the fossil record. Many of the fossils from this time show that they consisted of soft body parts. If you ever hear the term "Cambrian Explosion," this is what it is. Whereas before this natural selection played with a single can of Play-Doh, it appears to have stumbled upon an entire warehouse during the Cambrian Explosion. The forms that emerged during this time are amazing. It's as if nature finally learned what Play-Doh can do.

Nowhere on earth is this more evident than in a formation located in the Canadian Rockies called the Burgess Shale. It was discovered in 1909 by a very lucky paleontologist named Charles Walcott. He was about to call it quits for the summer and return home when he chipped away at an outcropping of black shale. What he found were fossils, but not the types of fossils anyone had ever seen before. By the time of his death in 1924, Walcott had discovered as many as sixty-five thousand different species of extinct animals.

What the Burgess Shale reveals is nature's laboratory. The life forms preserved there served as the prototypes for every life form we find on earth today. The variations in shape and forms are astounding in their abundance and complexity.

With this complexity and abundance, competition for resources was kicked into high gear. Organisms had to find alternative ways to compete with their oddly shaped neighbors. New body plans emerged. These new plans employed new methods of winning the race for resources. And then one day, a dark and gloomy day, a new organism emerged from the darkness. This creature wasn't content to compete for resources. It saw fit to devour the competition.

We have here the very first predator.

And life would never be the same.

Chapter Twenty-Eight

An Arms Race

What happened 542 million years ago to cause the explosion of life we see in the Cambrian Era?

It was a war.

Before we dive into the cause of that war, we must first ask ourselves why there seems to be more life created during the Cambrian than in previous eras. To be completely honest, that's only one of the questions begging for an answer. What we should also ask is why all those creatures present before the Cambrian Era are not reflected in the fossil record. The answer to that is simple. We've already alluded to it. They had soft bodies.

So what's wrong with soft bodies? Well, nothing to speak of, unless you are trying to make a fossil. Before the abundance of complex shapes we see in the Cambrian, soft bodies ruled.

Let's pause here for a moment and look at what it takes to make a fossil. It's not easy. You need an almost perfect storm of things to occur for the creation of a fossil. First, an animal has to die. That's the easy part. All animals die without exception. We can count on that much. As far as our fossil making machine is concerned, the animal needs to die in the right place, one where it can be buried in loose sediment. After it's thoroughly buried and protected from scavengers, it needs to mineralize. To mineralize, you need hard parts to work with. Being a worm with a soft body simply won't do. There's not much there for the fossil making machine to work with. After that mineralization has taken place, the sediment it is buried in has to harden and become rock.

Once all of this happens, you have a fossil. It then has to be found by some enterprising fossil hunter, a Mary Anning or Charles Darwin, and there were billions of years of geological upheavals and changes working against them. Fossils need to survive all this to be found millions of years after formation. It's actually quite a wonder we have found any at all.

There is one thing you can be sure won't be found. You won't find soft-bodied organisms in the fossil record. They don't lend themselves well to the perfect storm of events I've just outlined. They decay fast. The sediment obliterated them into nothingness. Even if they did happen to thwart decay and obliteration, they would still need to mineralize. The problem there is they don't have the hard body parts necessary to mineralize. For all intents and purposes, they are ghosts. We know they are there, we can sense their presence, but we can't see them.

I'm very fond of saying that "something happened." That's because things do happen, and we may not always know exactly why or how, but we usually know when and where. We do a pretty good job of theorizing about the "why." The "something happened" we are concerned with now is the one that happened during the Cambrian explosion—that marvelous epoch where life forms seemed to step out of the shadows and scream for attention. These life forms weren't soft-bodied worms or fragile multicellular organisms. Not at all. These creatures had shells. They had armor to protect their soft parts. They came ready-made for battle and protection. But, you might ask, protection from what?

Remember. At one point in time, in the pre-Cambrian, life existed below the earth's early oceans' surface. This life was often stationary. They stuck close to the bottom. They pulled the nutrition from the water around them or waited for smaller organisms to pass by. They loved to consume the unsuspecting passer-by. A few of these early life forms became mobile. They discovered it was better to move around to look for food than to sit around and wait for it. Those creatures that were mobile, or could steer themselves in the direction of food, or anywhere until they found food, had an advantage over those that couldn't. The creatures that served as their next meal needed to adapt to avoid being eaten. They needed to successfully hide in the shadows to avoid detection, or they needed to develop ways to fight back. In other words, an arms race had begun.

As Darwin said, "It is difficult to believe in the dreadful but quiet war lurking just below the serene facade of nature." For example, let's say that I'm a soft-bodied organism content to stick to the ocean bottom. For thousands of generations, my ancestors before me did the same thing. We had become experts at it. We knew if we waited long enough, perhaps moving now and then, food would come to us. It always did. And we were prolific when it came to reproducing. There were a lot of us.

On one unfortunate day, something emerged from the darkness. This something had teeth, and it could move fast. If we managed to escape its teeth, we still had to avoid the spikes that ran along its hard body plates. What it needed plates to protect it from we didn't know, nor did we want to. All we wanted to do was to survive. And that scary something wasn't the only one. There were others just like it. If we had any hope to live long enough to produce children, we needed to protect ourselves. Nature provided us with plates as well. We evolved hard external skeletons that successfully deflected the spikes. This made it slightly more difficult for the predators that came out of the darkness to consume us. Some of us developed tails to propel ourselves away from the predators. Unfortunately for us, the predators evolved as well. They grew armor-crushing claws and moved even faster.

It was a battle to the death.

The hard-shelled creatures are the arthropods. They appeared in great numbers, shapes, and sizes during the Cambrian explosion. As predators evolved to better catch their prey, their prey evolved unique and efficient ways to avoid getting caught. Because of this sudden arms race we see the proliferation of body forms that mark the Cambrian Era. These arthropods dominated the world for 420 million years. Predators such as the eurypterids, which looked a lot like giant sea scorpions, emerged during this time. The real winners were the trilobites. These small creatures were the most advanced form of life on the planet for 250 million years. Some of them could grow as large as three feet long. Some were as small as beetles. The trilobites enjoyed an exoskeleton, which gave them the advantage of protection.

Their exoskeleton did something else too. It helped keep the moisture in when a few brave little souls stuck their heads out of the water and discovered a whole new feeding ground.

Dry land.

There was no competition for food resources there. Even better, there were no predators to eat them.

At least not yet.

Chapter Twenty-Nine

Out of the Sea

THE SEA WAS FULL OF LIFE A HALF BILLION YEARS AGO. ARTHROPODS fought to survive. Present and abundant were worm-like organisms with soft skin. There were also creatures like the *Haikouicthys*. *Haikouichthys* was different from its neighbors, and it was fast. Fast enough to avoid predation. It also had something else the other creatures didn't. Something that differentiated them from the rest. An invisible secret. I say "invisible" because it was buried inside of them. It had a marvelously new structure that ran from its tail to its skull. Yes, I said skull. This also marked it as different. Skulls were new on the scene, as was this secret structure, called a notochord. These notochords served as backbones for *Haikouichthys* and others like them. The arthropods didn't have notochords. Remove their shell, and there wasn't much there. While shells were great for protection and hunting, they were also limiting when it came to size. They could grow, and their shell could grow, but it reached a point where it wasn't going to get much larger unless they could shed it to grow a new one, which left them soft, vulnerable, and without support during the process. These were invertebrates, meaning they lacked a backbone. These new organisms on the scene, the ones with the notochords that would soon develop into backbones and skeletons, are the first vertebrates. They weren't much different from fish, although they weren't like any fish you would see today.

Needless to say, they were here, and now that they were, they were here to stay.

While all these creatures struggled to survive in the sea, some exciting things happened on land as well. Remember the cyanobacteria—those first single-celled prokaryotes who started everything? They had a grand old time, as did their eukaryotic cousins, the algae. While the vertebrates and invertebrates fought beneath the waves, these precursors to plants lived the quiet life near the shores. As the tide moved in and out, some found themselves stranded on the moist sand and remained there, consuming nutrients, basking in the sun, and producing oxygen. This was a new environment. As we've learned, with new environments come new opportunities to thrive. They grew and slowly spread inland.

The world saw its first plants, and they spread like wildfire. Nothing was there to stop them. The world, at least the world of dry land, was their oyster. If things became too dry, they could rely on the rain to cool things off. It would also break down the nutrients in the soil for them to consume. Much as they'd done for the past few billion years.

As the saying goes, "All good things come to an end." When one group is having fun, there will always be another group who sees fit to spoil the party. In this case, it was the vertebrates. But we can't blame them entirely. After all, things in the ocean were getting pretty bad. They had predators making life difficult, if not horrible. They also had to compete with pretty much everyone else for food. The quiet life on land that the plants enjoyed must have looked pretty appealing to the first vertebrate who poked its head out of the water. As it supported itself with its new backbone and fins, it saw a new opportunity to escape the battle.

Of course, this wasn't done consciously. I've embellished this for dramatic effect. It's not like these early vertebrates thought, "You know what? Things are pretty bad down here. We're being eaten by sea scorpions, and there's never enough food to go around. When we do find something to munch on, we have to constantly look over our shoulders. Things on land look pretty quiet. It's time to pack our bags and go."

Remember, the wizard behind the curtain is natural selection. When you have large predators like the sea scorpion patrolling the deeper waters, those animals who can venture into the shallower waters for more extended periods have an advantage. And those who can stick their heads out of the water have it even better. This allows them to move into the shallower waters. The sea scorpions, or pretty much any predator to speak of, won't be able to reach them. It's quiet in the shallows. One of the things these adventurous animals had to do next was to breathe air. They still required oxygen, and their gills only worked when wet. They can't absorb oxygen from the air. So new methods of absorbing oxygen were needed.

A perfect example of this is the velvet worm or, at least, the ancient ancestor of today's velvet worms. Sitting somewhere between the worm and insect worlds, this creature lived in the sea—at first. As it ventured closer and closer to land, it developed tiny holes on its sides to breathe with. This allowed the worms to spend short periods on land. They were one of the first creatures to do so a half billion years ago. And because they had soft skin, they needed to remain in moist environments to keep from drying out. The small tributaries that stretched inland were shallow enough to prevent the predators from following. For the creatures that could follow these tributaries and venture out of the water for short periods, this must have been Nirvana.

Don't forget, nature doesn't play favorites. If the worms could make the giant evolutionary leap onto land, you know the predatorial arthropods wouldn't be far behind.

One thing that the arthropods' hard shells were very good at was preventing their skin from drying out. They helped to retain moisture. When they eventually evolved the ability to move into the shallows to feed and were able to support themselves on their many jointed appendages, the time for peace had ended. After all, there was a war on, and, in this war, there are no winners. There's only survival and reproduction. Do that, and do that well, and you can stay slightly ahead of the other guy. But he will catch up. He always does. Nature sees to that.

Developing methods to breathe the air was like a one-way ticket for most creatures. Once they could do that, they couldn't go back. The sea was no longer their home. If they tried to retreat into it, they would drown.

If you were to step back in time, you would see some amazing things crawling out of the water to gulp at the air. There would be worm-like creatures, amphibians with moist skins, and hard-shelled crustacean-like animals. The vertebrates had a much better time of it. Remember, they had developed inner skeletons that allowed them to support themselves better and to fight against this new thing they encountered called gravity. They were also able to move faster and to grow larger. They spread inland. Away from the predators until they began, of course, to eat each other.

You might also have seen something else. Something that wasn't a worm or a crustacean. Remember, I mentioned fish-like creatures with fins. They weren't content to watch everyone else move into the brave new world of dry land. It was their turn. How this happened is up for debate, but there are many reasons why it might have. Don't forget, tributaries stretched inland. There were also droughts. Venture into these tributaries or rivers during a drought, and you may end up trapped. To survive extended periods in the shallows, these creatures would be required to develop the ability to breathe the air. Then again, the simple fact that their food source walked out of the water might have pushed them to try it themselves. Any modification that allowed them to chase their food onto dry land would extend to their offspring.

The fact remains that around 3.5 million years ago, one of these fish-like creatures took that step, and it was a big one.

And how do we know about this creature? Because we found it. Or, I should say, American paleontologist Neil Shubin found it. In 2004 Shubin and his team found the fossilized remains of this transitional creature on Ellesmere Island in Northern Canada. I say transitional because that is exactly what they were. Shubin and his team named the fossil *Tiktaalik*, and it represented a species that was half-fish and half-amphibian. Consider them the first tetrapod, or four-legged animal. From them came everything else. And I do mean everything.

We've had quite a time of it since our early days. In the last four hundred million years, we have spread to the far corners of the earth. It's been an incredible ride. Some of us even ventured back into the ocean after living an extended time on dry land. Seals have done this. Whales have as well, and their closest cousin is today's hippo. At some point in the distant past, a hippo-like creature, a common ancestor to today's hippo and whale called the anthracothere, decided it had had enough of dry land and slowly moved back into the sea. Whales still have the remains of legs buried in their blubber. They just don't need them anymore. They also still have their lungs and, because of this, need to come up for air.

You may be a bit fixated on something I said at the beginning of the previous paragraph. I said, "we have spread" after introducing the fish-like creature discovered by Professor Shubin and his team. That wasn't a mistake. By "we" I do mean to imply that we are related to those early transitional creatures.

Let's find out how.

Chapter Thirty

Is Everything Related?

All stories are written using letters. The letters may differ somewhat depending on the language, but there would be no story to tell without them.

The story I want to tell you about now is no different. It's all about letters—four letters, in fact. The English alphabet has twenty-six letters. Twenty-six marvelous letters that have combined to create many of the great literary works we enjoy as well as the words I am writing right now. Those twenty-six letters are pretty powerful when you think about it. They have moved men to war and have also inspired some incredible creations. Some of which have enabled us to scan the depths of the ocean and walk on the moon.

But this isn't about our language of twenty-six letters. It's about an even more powerful language comprised of just four. Four simple letters that, without them, there would be nothing at all. At least as far as life on earth is concerned. Perhaps all life. Everywhere. This language of four may have come from somewhere else for all we know. But it's here now. And when you string those letters together in different ways, a miraculous story is told.

What are these four letters?

A, C, G, and T.

That's it. Three consonants and a vowel. That's the translation we work with. Like one would translate a papyrus in ancient Egyptian into English, we use letters we are familiar with. In this case, A, C, G, and T.

The title of this story, if you haven't guessed, is DNA.

DNA stands for deoxyribonucleic acid. That title, unabbreviated, would scare most people away. DNA rolls off the tongue. It's easy to say.

Open its cover to look at the table of contents, and you will see a chapter on every creature and organism that has ever lived. Flip to humanity's chapter, and you will see those same four letters, A, C, G, and T combined and repeated three billion times. It's in the way they are combined that tells the real story.

Think of it as a recipe, or instruction manual, for life.

I should mention briefly what those four letters stand for. Each letter represents an organic molecule. Scientists call these molecules nucleotides. The "A" stands for Adenine, the "C" is Cytosine, the "G" is Guanine, and the "T" is Thymine. They make up what is called the genetic code. It is the instruction manual for building all living organisms.

The code used to build you is the same code that made me. Compare yours to mine, and it's 99.9 percent identical. If that last.1 percent were to match, we would be twins. Or clones. The code takes proteins and puts them together in beautiful and elegant ways. If you were to look closely at humanity's chapter, you would see twenty-three sections to that chapter. Each of those sections are called chromosomes. These sections contain everything there is to know about building a human. They tell cells how to put proteins together. Each step in the instructions we can call a gene.

We learned about this in 1953. That was when James Watson, Francis Crick, and Rosalind Franklin discovered the structure of DNA. The structure of life's instruction manual. The question they asked was how it all worked? How was the information to create a mouse, or a human being, stored?

Much has happened in the years that have passed since that discovery. We know a lot more about that instruction manual. We know humanity's chapter contains three billion letters. The Human Genome Project, established in the 1990s to map the human gene sequence, has helped us understand the lines of this chapter. We also know some other things. We know if you took all of the DNA in a human body, separated each strand, and line them up, one after the other, we could create a line of DNA that stretched from here to the sun and back. Not once but six hundred times.

We also know if you gave that chapter to a typist and asked them to type it out, it would take fifty years to write. That's if the typist put in a standard eight-hour day. Maybe longer if they took an hour for lunch.

How much would we have at the end of those fifty years? At the standard 250 words per page, our tired typist would have produced one million pages of text.

The Human Genome Project achieved just that, and took thirteen years to complete. Hundreds of scientists from all over the world were involved, as well as several anonymous volunteers who donated their blood and DNA. As impressive as the project's mapping of the human genome is the story that it tells when you begin to compare it with other chapters in the book of life.

This is where things get even more interesting. In the late 18th and early 19th centuries, naturalists like French biologist Georges Cuvier compared the anatomy of living creatures to determine their classifications and chart their differences. Today, geneticists do the same thing but at a deeper level. They look at our genes, the lines of the instruction manual, and compare those lines to those in the other chapters found in the book of DNA.

Here, the true nature of the story, the *real* story of life, comes out. Like any good book, each chapter must relate to one another. Turn to the chapter on mice, and you will find that we are 99 percent the same at a genetic level. It is why lab mice are as popular as they are. How things affect them will most likely, at least 99 percent of the time, affect us. It's that 1 percent that we have to be careful about.

What about man's best friend? Flip to the dog chapter with its German Shepherds, beagles, and retrievers, and we find we are 94 percent the same. That's a 94 percent genetic similarity with the animal many of us share our homes with.

Years ago, when I began to study evolution, I sent an email to molecular biologist Sean B. Carroll. I had just read his book *Endless Forms Most Beautiful* and wanted to know if we shared any of our genes with plants. Did our genetic similarities end with animals, or did it extend to other living things on this planet? His answer was "yes." Pick up a grain of rice, for example. We share 25 percent of our genes with that little guy. Our common ancestor, the one we shared with that bit of rice, existed 1.6 billion years ago. Since then, we have gone on our merry way. Look at a pumpkin and you are looking at a living thing that shares 75 percent of our genes. A sea sponge? 70 percent.

And if you are wondering about gene compatibility after all this time, consider this: Scientists have been able to place human genes into a fruit fly's genome, and they work just fine.

We all came from the same place. Stand next to an oak tree, and you are standing next to a distant, very distant, cousin.

Our closest relative of all is the chimpanzee. Some estimates place the genetic similarities with our fellow primates at 99.4 percent. We are not much different, after all.

Over 220 years ago, Charles Darwin's grandfather Erasmus Darwin saw these similarities in all living things. Everything around us, he felt, had to be related in some way. His conundrum was that he didn't understand how. But the similarities were there.

As the earth and ocean were probably peopled with vegetable productions long before the existence of animals; and many families of these animals long before other families of them.

He looked toward the sea for answers, as many thinkers had before him. In plants, there was beauty to be found. The same beauty seen in birds, in the trees they built their nests in, and the worms in the earth. We are all put together a little differently.

Shall we conjecture that one and the same kind of living filaments is and has been the cause of all organic life?

Four billion years ago, we all parted from our last common ancestor. Since then, the book of life has been busy producing new pages and chapters. Some of those pages have faded over the years. They've reached the point where they can't be read at all. The marvelous organisms that looked out from those pages are long gone. By reading the pages we have, we can piece together the missing ones.

One thing we do find is that even the best and most accurate typist makes mistakes. As we shall see.

Chapter Thirty-One

Mistakes Were Made

For years people hadn't been quite able to figure out how snakes move. We have a pretty good idea of how they do it, but there is still some mystery. Snakes hold other mysteries too. Like legs.

Yes. Snakes have legs. Not like the legs you or I have. Not anymore. And not all snakes either. Again—not anymore. But if you were to coerce a python or a boa constrictor to sit in front of an X-ray machine, you would see two tiny legs buried near their tails and hidden by muscle.

But why? They don't need them, and they certainly serve no purpose anyone can see, so why are they there?

It's because 100 million years ago, a group of lizards developed with shorter legs. These shorter legs allowed them to remain close to the ground, perhaps hidden from predators in short grass while they searched for food. Their fellow lizards, those with larger legs, were easy to spot, so they were eaten.

Flash forward a few million years, and those legs have all but disappeared as the lizards slithered closer and closer to the ground. They became quite adept at it—and fast. So today, they appear to be gone. But vestiges of them remain. It is why we call them vestigial features. Remnants of past features long gone and no longer needed.

We have them too. Not hidden legs, but there is an organ we can point to that apparently serves no purpose. It's called your appendix.

You will read different things about it, but the bottom line is that you do not need it. Not anymore, anyway. Your ancestors probably did.

Charles Darwin also famously weighed in on its purpose when he suggested it may have been used by an ancient ancestor of ours to digest plants and leaves. Some mammals today have a structure very similar to our appendix, called the cecum, which is used for this purpose.

So why does this happen? If you remember the chapter on Jean Baptiste Lamarck, he proposed that physical traits come and go because of a system driven by "use and disuse." He felt some physical features, if no longer needed, faded away. Like the snake's legs. Lamarck would tell us they no longer served a purpose, so they eventually disappeared. What Lamarck missed was that somewhere along the line, a lizard was born with smaller legs. While it was a birth defect, that happy development ultimately proved advantageous to the lizard and its offspring. It allowed it to escape long enough to reproduce, producing more lizards with shorter legs. It wasn't "disuse." It was a mutation.

In the X-Men movies, the superhero Wolverine and the rest of the team were born with their powers. Wolverine's powers may have been augmented by adamantium, but he was born with the ability to heal, which is far and above anything we can do. And don't forget his retractable claws. The X-Men are mutants. Defects in their genes brought out features and abilities that were not present in the population before their births—defects that enhanced their chances of survival.

Wolverine or Storm didn't choose their mutations. Nor did their parents. They occurred during the copying process when their cells divided. Mistakes were made during DNA replications that proved to be useful. It could have gone another way. It most often does. Most defects result in an organism's inability to survive in the environment it is born into.

In the case of our appendix, a series of defects rendered it useless. Fortunately for us, these defects proved to be neutral when it came to our chances to survive. With snakes, the defects that reduced their leg size proved to be advantageous. And what of the dodo bird? That long-extinct bird with wings it could no longer use? The defects that led to its inability to fly weren't exactly disadvantageous until it was presented with the dangers that led to its extinction.

Naturally occurring mutations happen every day. Invisible little mistakes in replication. Not all mutations are the result of accidents in DNA replication. Some happen because of DNA damage.

Radiation is an excellent example of an outside force that can damage DNA. Chemical spills are another. Organisms that encounter these spills can suffer damage to their DNA, which is then passed on to their offspring as a defect—or at least a defect when compared to the original DNA.

In the last chapter, we looked at the book of life contained in DNA and the four letters that, when combined in marvelous and unique ways, give rise to every organic thing around us. Every once in a while, a page might not be copied correctly or is left out entirely in the copying process. Over 90 percent of the time, when you come across one of these books, you'll toss it out. Less than 10 percent of the time, these missing pages or misplaced sections won't be missed. Perhaps it doesn't move the story along, and you don't notice it. So you keep the book. Every time it is copied, those pages will no longer be there. No harm, no foul.

What if the missing page or misplaced section actually helps the story? You can imagine this would be a rare occurrence, but it can be magic when it happens. So magical that all future copies are preferred over the original, complete book. The original book is tossed aside, and the defective copy survives.

Some features shine a spotlight on the fact that we sometimes miss a page during editing.

Let me present to you the laryngeal nerve and the vas deferens. Both are found in humans, and the vas deferens is only found in men.

Let's start with the laryngeal nerve. Traveling from our brains are cranial nerves. One of these nerves branches off to both the heart and the larynx. The larynx is your voice box. Two cranial nerves connect to it. Think of them as internet cables that transfer information. They tell the larynx what to do. I mentioned that two of these cables, or nerves, go from the brain to the larynx. One of these travels from the brain to the voice box in a straight line. A nice short cable optimized for signal transfer. If you designed it yourself, you would do the same thing. But the other nerve appears to have been installed by someone who wasn't quite sure what they were doing. That nerve heads down to the chest first before traveling back up to the larynx—a nice long route for no apparent reason other than it was designed by nobody.

Which leads me to the vas deferens. This little tube connects the testes to the penis. As the testes produce sperm, they send it off by way of the vas deferens commuter line. The distance between the two isn't very much. The tube only needs to be a few inches long. Instead, it travels away from the penis before looping back. It would be like sending a train from Virginia to Florida by way of Boston. If you happened to be on that train, finding yourself heading north instead of south, you would complain. Mistakes had to have been made, right?

And did you know you have a blind spot?

When light hits your eye, it stimulates the retina at the back, which is covered with light-sensitive proteins. The information created by the stimulated proteins is carried back to the brain by the optic nerve. The thing is, the optic nerve is plugged directly into the retina. Because of this connection, there are no light-sensing proteins at that spot. So the information that gets transmitted back to the brain has a little blind spot. Our brains have become so used to this that they compensate for the blind spot by filling it in. In the end, you aren't even aware of it.

The things our brains have to put up with.

Chapter Thirty-Two

A Whale of a Tale

In the last chapter, I touched on vestigial features, like the leg bones in some snakes and the human appendix. Most of the chapter was focused on the apparent mistakes in design brought about as animals evolved. I was reluctant to use the term *mistakes* because it implies purpose. When it comes to mutations, adaptations, and evolution, purpose has no role to play.

If all fossil evidence of evolution were suddenly wiped out tomorrow, we would still be surrounded by other pieces of evidence. Living and breathing evidence.

Like whales.

At one point in the far distant past, an ancient ancestor of ours left the ocean to crawl onto dry land, perhaps to search for food or escape the large predators that wanted to make a meal of it. Either way, this ancestor gradually gained the ability to exist in shallow waters, which allowed it to avoid predation and find food where there was less competition. It lived long enough to pass these traits on to its offspring. Generations later, they made the move from the ocean onto dry land. They then spread out, encountered different environments, and evolved into the many beautiful creatures that populate this planet—including us.

But what about the denizens of the ocean? They also continued to evolve as the aquatic arms race picked up steam. The same thing happened on land. It was a battle for resources. A struggle that drives evolution and sometimes kicks it into high gear. This never-ending war may have caused some land-dwelling creatures to gaze longingly back at the ocean, forgetting why they left it in the first place. The grass isn't always greener, especially in nature.

I'm taking poetic license here. Most creatures do not gaze longingly at anything. One group of animals, perhaps a long-extinct four-legged animal called *Pakicetus*, found that the safest place to avoid the predators it faced on land was to stick close to shallow water. When a predator lunged out of the tall grass, *Pakicetus* was safely out of reach. Some predators refuse to get their feet wet.

Eventually, our little *Pakicetus* began to evolve. Remember, evolution works over millions of years. In time, *Pakicetus* gave way to *Ambulocetus*, a group of creatures given to spending a lot of their time in the water. They could swim. They enjoyed the water so much they stayed, evolving into *Kutchicetus*. Next up was *Dorudon*, whose nasal cavity had moved toward the back of the head to allow for extended periods underwater, and then, about five million years later . . .

The whale.

When you look closely at a whale skeleton, you can still find its hind legs. They are small and useless appendages located precisely where you would expect to find hind legs. Vestigial features inherited from an ancient relative. A genetic mutation caused an ancestor's legs to develop differently, and that difference didn't hinder their chances of survival. In fact, it may have helped. So it was passed on, just like that nasal cavity that gradually moved from the front of the head to the top and back. It became what we call a blowhole. With each mutation that moved the nasal cavity, the animal who suffered the mutation found it could keep its head below water for longer periods. More time gave it extended opportunities to look for food or to keep an eye out for predators that may attempt a sneak attack from below. There are many other examples of vestigial organs in the animal kingdom, like the mole-rat. These little rodents are blind, and it's not because they don't have eyes. They do. But their eyes are now vestigial organs. They no longer use them. They couldn't even if they wanted to. Their eyes are entirely covered by a thin layer of skin.

Charles Darwin was fascinated by vestigial organs. Especially those that could be found when he looked into the mirror. He touched on this in his 1871 book *The Descent of Man*. Things like wisdom teeth, or the muscles of the ear.

There are more. Like goosebumps. Do you know what else gets goosebumps? Porcupines do. That's what causes their quills to go up when they are frightened or alarmed. The same thing goes for cats and dogs. Their hair raises for the same reason. It's a fear response. It functioned as a mode of self-defense to make the animal appear a little bit larger when encountering a predator. Or, when it's cold, it helps to trap air, warm it, and serve as a sort of virtual blanket.

That's what our goosebumps are doing. There isn't thick hair there for the goosebumps to make rise anymore. Natural selection saw to that. But the reaction remains.

And then there's the ability to produce vitamin C, an ability most animals have. Except us and a few others like some species of bats. Guinea pigs can't produce it either. The gene that creates the enzyme to produce vitamin C was shut off a long time ago. It was disabled by natural selection. When it's working, like it does in most animals, vitamin C is produced. For some animals, like us, it's an important nutrient. A very important one. When the gene is working, there is no need to find food that contains it. When it's not, then it's essential to do so. We have the gene still. It can still be found in our DNA. It's just been turned off—further evidence of an evolutionary change that would have proven disastrous if we could not find other means to bring vitamin C into our bodies.

This leads to my favorite example. An example that was extremely painful for me when I tried snowboarding for the first time a few years ago. I landed on it one too many times and couldn't sit for a week. We call this our coccyx. It's actually a tail, one that natural selection shortened and basically buried in our backside. Like the whale's hind legs, we can no longer use or see them.

Our DNA contains traces of our past. There are switches in our genes that have either been shut off or turned on over the years, while natural selection's fingers hovered over the controls. I know I'm anthropomorphizing natural selection, but what's wrong with a little metaphorical fun?

Chapter Thirty-Three

Putting the Selection in Sex

For much of this book, I've focused on evolution by natural selection. Along the way, we've encountered thinkers from the past who either got there before Darwin with little fanfare or who contributed to the idea. I've also touched upon reproduction but have left it pretty much alone, assuming the reader is fully aware of how it works. In a nutshell, sexual reproduction allows an organism to combine genes during a new organism's formation. In simple terms, the new organism receives half of both parents' genes, mixes them up, and becomes something new. Not necessarily something new as in a new species, but something new as far as which genes express themselves and which features appear. During this swapping and mixing process, a mutation might occur that throws a small wrench into the works, creating an unexpected feature like a bulbous nose or six toes instead of five.

Reproduction gives natural selection something to work with. If everything were the same, as in the case of an exact clone, there would be no advantages to be had. In a colony of duplicates, where all organisms are the same, a sudden change in the environment would wipe out the entire population if that change was a detriment to survival. Let's say there is a sudden cold snap for which the organisms are wholly unprepared. If a small percentage of those individuals had small amounts of hair that helped them last longer than the others, natural selection might step in to hand out survival awards.

Without reproduction and the swapping of genes, there would not be anything around us. There wouldn't even be an "us." There would be nothing: no grass, plants, trees, or fish. There would be no fossils to be found, nor would there be any one or any *thing* to find them. Life wouldn't have survived the very first ice age. Dinosaurs wouldn't have ruled the world, and when the asteroid that wiped them out hit the earth, it would have hit a barren, lifeless planet. There would have been no extinction event since there would not have been anything here to go extinct.

There are pressures on individuals to mate. The entire fate of life on this world depends on it. On top of the competition for resources, there is a competition for organisms to reproduce. And those who have an advantage when it comes to finding mates have a better chance of producing offspring that have better chances of finding mates. It's a game. A game where the organism that has the most sex wins.

For Darwin, the peacock's tail was a conundrum. Here was an organism that did everything in its power *not* to remain hidden from predators. Especially the male of the species. It is almost as if it proudly announces its presence on the food chain. The female of the species, on the other hand, does not have the ornate tail feathers displayed by the males. Their feathers are dull in contrast. It took Darwin some time to figure this out. He knew there had to be an advantage for having these feathers; otherwise, they wouldn't be present, and we wouldn't be talking about them.

What if, he wondered, the feathers served some purpose in mating? What if the male with the flashier plumage attracted more females? And perhaps the females, whose job it is to protect their eggs and young, have a duller set of feathers to not be so obvious to predators. Perhaps the female selected the male with the best display. It would make sense that the male with the most ornate display would produce offspring with more ornate displays if they were to mate more than the others. Over time the displays would become more profound. All because of a simple selection process.

Sexual selection.

Darwin would eventually divide sexual selection into two areas. There was the area defined by the peacock's tail—that of aesthetic value, or the ability to attract a mate based on subtle, or not so subtle, cues in appearance. And then there was the area of combat, where two males fought over a female. Like the bighorn sheep. These sheep have huge horns, which they use to ram each other. It's not because of some disagreement. It's because they are fighting over a female.

Sexual selection isn't confined to the same species either. Flowers have developed over time to attract bees. When a bee lands on one, they are coated with a cloud of fine pollen dust which they transfer to another flower the next time they land. Unbeknownst to the bees, they are unconsciously selecting for the more attractive flowers when they land upon them and transfer their pollen. Over time the population of flowers that display those traits that the bees are attracted to expands.

For Darwin, the idea of sexual selection explained a lot of what he saw in the animal kingdom. And it wasn't just the animal kingdom. He viewed the wearing of jewelry and makeup as nothing more than another tool in the arsenal to attract human mates. Males fought over females, and females did everything they could to attract the males who fought over them. Sexual selection could be as much a deciding factor for the presence of certain traits in a population as natural selection was. He wrote as much in his book *The Descent of Man and Selection in Relation to Sex*. There are no ambiguities there.

But not everyone agreed with this assessment. Even Alfred Russel Wallace, who supported Darwin in almost all his ideas, couldn't exactly buy into this one:

In this extension of sexual selection to include the action of female choice or preference, and in the attempt to give to that choice such wide-reaching effects, I am unable to follow him more than a very little way; and I will now state some of the reasons why I think his views are unsound.

Wallace felt female choice among animals was going a little too far. Animals couldn't understand beauty, not in the way humans can. To say the aesthetic values of pretty plumage or a fancy dance could attract a female was giving animals with low cognitive abilities too much credit.

While Darwin gave sexual selection as much importance as natural selection, Wallace countered it should be restricted to combating males.

While they agreed the fight for mating rights was a form of sexual selection, the more subtle choice of a mate remained a point of contention. For Wallace, sexual selection does have a place in adaptation, but it is only another form of natural selection.

I prefer to look at it this way. Both men are right. Darwin was right in that selection of an organism by another organism is a powerful mechanism for natural selection. It is on equal footing with adaptations that allow an organism to survive in an environment, like a longer neck to reach food where there is less competition or coloring that enables an organism to draw a mate's attention. In the latter instance, the environment they adapt to is one of competition.

Let me put it another way. Wallace's assertion that sexual selection is not a separate process, as Darwin preferred, but another aspect of natural selection, is right on the mark as well. Wallace's mistake was in thinking that an organism's preference for one mate over another had something to do with aesthetics, at least as we commonly think of the term.

A bee can't admire the Mona Lisa for its aesthetic value as far as we know, but it can develop an affinity for bright orange flowers of a particular shape. Not because they find it is pleasing but because nature selects for bees attracted to orange flowers, and the orange blossoms are selected by nature because it attracts bees. There's a mutual benefit in those adaptations. These selections are not made consciously. They are hardwired into the nature of things.

Remember, it's all about the perpetuation of genes. How a gene, or set of genes, survives depends solely on the organism's ability to survive long enough to reproduce, to pass along copies of these genes. Be it due to the organism's ability to withstand extreme temperatures, avoid predators, find food, or even attract a mate, it is natural selection at work.

But what if there are other factors at work that seem to make no sense at all? What I've just described is a selection process involving selfish organisms and their genes. What about situations where selfishness is absent? How does natural selection account for altruism?

Chapter Thirty-Four

Altruism

There are a lot of squirrels in my backyard and many things they are afraid of. When a fox comes around, it might be a good idea to sound an alarm, but it's not quite a good thing for the squirrel who raises the alarm. If the fox didn't know he was there before the warning cry, she does now.

This doesn't only happen here in my backyard. You can find similar acts throughout the animal kingdom. In Africa, vervet monkeys cry out when a predator approaches. It has been estimated that they have thirty different alarm calls. Each call identifies a different type of predator. The warning that says a leopard is approaching is much different from the one given if you or I were approaching.

Sticking to the world of primates, we can also look at bonobos, who have been seen helping sick or injured companions. Chimpanzees share food. It doesn't stop there. Dolphins help injured or tired dolphins by swimming beneath them to push them to the surface to help them breathe.

Another example is vampire bats. They aren't the selfish little creatures we may think them to be. They form buddy systems and will regurgitate captured blood to feed sick bats.

There is clearly something going on here, and it seems to fly in the face of natural selection. If you walk down the street and see a child fall off of her bike, you will stop to help. That's not much different from what we see in the animal kingdom. We see warning calls, the sharing of food, and grooming. That's not a battle for survival at an individual level. It's cooperation. Another word for it is "altruism," and it was coined by French philosopher Auguste Comte. He was asking us to "live for others." He envisioned a world where we would set selfishness aside and cooperate with one another.

Doesn't this contradict the whole notion of survival based on selfishness? That's what Richard Dawkins said in his landmark book *The Selfish Gene* in 1976. Genes are selfish. The term does seem to imply a motive on the part of the genes, and Dawkins is anthropomorphizing them to illustrate a point. The gene isn't necessarily striving to survive; it is just surviving long enough to be copied. The genes that are successful at doing this over and over are the ones that will continue. Those that are not successful disappear. Their race is run.

So where does altruism come in? Why does a vampire bat regurgitate its food to feed a sick roost mate? It is certainly of no benefit to the regurgitating bat. In the case of the warning calls, the benefit is completely for those who don't see the predator coming. The one giving the warning is the one in trouble. All he had to do was keep his mouth shut.

So why didn't he? Why not sit quietly, protecting his genes, while the predator slowly creeps? Even if you're skeptical and are of the mind there is no intent there, the warning call is a reaction to danger. There is still an action or a reaction that benefits others and not the organism itself. That can still be considered altruism. There are positive consequences to that action.

Charles Darwin sought to explain this. Being the one who delivered the concept of natural selection to the world stage, he understood the contradictions that altruistic behavior brought to the subject. It needed an explanation. You can't have both. There had to be some sort of evolutionary advantage, or cause, behind cooperation.

This is what troubled him, and he wrote as much in his 1871 book *The Descent of Man*:

He who was ready to sacrifice his life, as many a savage has been, rather than betray his comrades, would often leave no offspring to inherit his noble nature.

If a noble creature sacrifices its life for its companions, then those noble genes are sacrificed with it. Darwin didn't know of genes or genetics when he wrote his books, but he had perfectly captured the dilemma. Suppose there is an advantage to such behavior. In that case, it doesn't seem to make sense in the greater scheme of things because that self-sacrificing behavior would eventually be weeded out of the population. If my genes tell me to help you out, and in doing so I die, then I can't pass them on.

The solution Darwin came up with is group selection.

A tribe including many members who . . . were always ready to give aid to each other and sacrifice themselves for the common good, would be victorious over most other tribes; and this would be natural selection.

The idea is that natural selection not only acts on an individual level but also on a group level. This does seem to make sense. Think of an organism itself. By its very nature, it is a collection of cells that cooperate and perform as a group to survive in whatever environment they find themselves in. The better adapted the organism is, the more chance it has to pass those adaptations on.

There can be no denying that altruism exists. We see it every day. Every act of kindness can be used as evidence against the concept of a selfish gene. It doesn't matter what the actual intent is behind that act of kindness or cooperation. It is still something that helps to perpetuate one set of genes over another.

It does appear to be a conundrum. To highlight this, let's stick with the example I started with—the squirrel who cries out a warning when it sees a fox. Let's say that this is the only squirrel in the whole history of squirrels to ever do so. It alone is capable of formulating the need to warn others when a fox pops up. Let's also say it is very successful at this. So much so that the population of squirrels in the area grows and grows. They can avoid foxes and reproduce, all because of this one squirrel. This super-squirrel is so busy playing the rodent version of Paul Revere that it never settles down long enough to start a family itself. One day, tired from its constant vigil, it doesn't see the fox sneaking up behind it, and it gets eaten.

Those genes, those super-squirrel genes, are gone forever.

Where, Darwin and others asked, is the advantage in that? It should be a one-time event and a short-lived one. We shouldn't see altruism on the level we see it around us if the whole idea behind natural selection was every man or beast for himself.

A century after Darwin first outlined this problem in *The Descent of Man*, biologist E. O. Wilson had an idea as to what was going on. His 1975 book *Sociobiology* outlined an alternative to the idea that altruistic genes would be quickly weeded out. Altruism is essential for survival.

And it starts with our closest relatives.

Enter kin selection.

If somewhere in the distant past a genetic mutation caused an altruistic trait to persist, it only makes sense that the trait would be shared by all the offspring of the parent organism, mathematically speaking. This small group of organisms would have an advantage, and more opportunities, over organisms that didn't have this gene. So the gene would persist.

There's a part two to this as well. While kin selection accounts for altruistic behaviors between relatives, what about those outside of the immediate family?

At face value, it does appear to echo the golden rule. Treat others as you would have them treat you. But it is more complicated than that. With kin selection, we are programmed to help those closest to us. Reciprocal altruism falls into a grayer area. Here, unrelated members of a group help one another out. We've all seen videos of monkeys who turn their backs to one another so their companion can pick out flies. They do not have to be related for this activity to take place, yet they are doing so just the same. Is it because they are kind at heart, or is it because of kin selection that empathy and sympathy have found their way into our genes and are now used in a broader sense? There could even be a bit of "what's in it for me" thrown in.

Reciprocal altruism can be boiled down to a game. How does one benefit the most by displaying kindness or cooperation? It's a twisted form of the golden rule. At the level of the gene, it doesn't care, so long as that activity enables it to last long enough to replicate.

Whatever the reason, it is clear that altruism is here to stay, and somewhere along the line, as we evolved, those groups or individuals that cooperated passed these cooperative genes on.

That's a world I'm happy to live in.

Chapter Thirty-Five

Coevolution

A century ago, biplanes sped across the skies of Europe during World War I. Manfred von Richthofen, also known as "the Red Baron," piloted one of those planes. Richthofen has been credited with at least eighty victories in the air. How did he do this? With an MG gun barrel bolted to the front of his cockpit. Before this, pilots would point pistols at one another, hoping for a lucky hit. When gun placements became possible, they faced a very unique problem. With a spinning propeller, how does one shoot through it? One early solution was to strengthen the blades. It didn't prove to be a great idea, especially when some pilots shot their propeller off or had bullets bounce off in all directions. As a solution, engineers developed specialized gear to synchronize the blade timing and the firing mechanism. This way, the bullets passed harmlessly through the propeller. Harmless for the pilot, that is. As dogfights became more and more commonplace over war-torn battlefields, so too did the improvements.

Today there's another battle taking place over our heads. One not so different than the dogfights of yesteryear. Only this time, the part of the Red Baron is played by a bat. His intended victim?

A moth.

As the bat employs echolocation to locate the moths, for dinner, of course, the moth employs several defensive maneuvers. In some cases, these maneuvers involve a series of dives or loops to avoid the bat. In the case of the *Bertholdia trigonia*, a species of moth in the southwestern

United States, there is a defensive maneuver that even the Red Baron would have been impressed by had echolocation and sonar been a thing known to us in his day. What this little moth does when it detects a bat's attempts to find it using echolocation is fire off short bursts of ultrasonic noise with its wings. These bursts essentially jam the bat's radar.

So now what? Now it's up to the bats to come up with a solution.

Now, here, you see, it takes all the running you can do, to keep in the same place.

That quote comes to us by way of Lewis Carroll and his book *Through the Looking-Glass*. It was one of two books that chronicled the adventures of Alice. That other book is *Alice in Wonderland*. The speaker is the Red Queen and what she's telling Alice is some disturbing news. She's explaining that, in her world, you need to run faster and faster to stay in the same place.

She might as well be talking to our little friend the *Bertholdia trigonia* moth. As bats evolve new echolocation methods, the moth will need to develop new ways to jam it to survive. In this case, keeping in the same place means surviving to fly another day.

What I've just described with the case of the bat and the moth, and even, in a non-biological sense, the Red Baron's gun placements, is coevolution.

Think of it like this. It all goes back to our genes. In this case, it's the genes of two separate species battling for survival. An evolutionary change in one species may stimulate a change in another species. This goes back and forth and, over time, both species experience change.

So, just like with our bat and our moth, one change begets another. If the bat develops a slightly altered echolocation method, it will be the moth's turn to step things up. For this to happen, a gene has to change. Genes change all of the time, but many moths will have to die for the right gene to change if it ever does. It's not like the moths have any control over this. Suppose a gene change helps the moth avoid becoming an evening snack long enough to reproduce. In that case, that new gene might propagate through the population.

For the moth, the right change could mean it stays in the same place, the place in question being survival, or it falls behind and loses the race.

If the change in one organism is linked to a change in another organism, genetically speaking, coevolution is said to have occurred. The term itself was first brought to life in 1964 in an article written by Paul Ehrlich and Peter Raven. The article was titled "Butterflies and Plants: a Study in Coevolution," and was published by the *Society for the Study of Evolution*. In that title, with the introduction of that one new word, "coevolution," a whole new area of research opened up to evolutionary biologists.

It didn't take them long to jump in it either. Seven years later, evolutionary biologist Leigh Van Valen came up with the Red Queen Hypothesis.

Van Valen proposed that coevolution often involves an arms race. You have a predator and its prey who both up the game along the way, like the bat and the moth. Each one tries to outdo the other. Another example of this predator/prey game involves the common garter snake.

The West Coast's garter snake's diet includes the rough-skinned newt. The newts have a very unique defense mechanism to avoid coming to a gruesome end. They've evolved an extremely powerful neurotoxin that collects on their skin. In turn, garter snakes have evolved a resistance to this neurotoxin. So what did the newts do? They turned the toxicity up a notch. Many dead snakes later, another mutation allowed for greater resistance to the toxin. It's become a constant back and forth between the two of them. The newts become more toxic, and the snakes use the toxin as added spice to their meal.

It doesn't have to be an arms race. Coevolution between species can also be cooperative. In these cases, it's a mutualistic relationship.

If you've ever seen a spider crab with algae on its back, then you've witnessed mutualistic coevolution in action. The algae allow the fortunate crab to blend in with the ocean floor, which will enable it to escape the searching eyes of predators. In turn, the algae enjoy life on a spider crab's back. They have become allies in the battle for survival.

You and I have the same sort of thing going on inside of us. We don't have algae growing on our backs, but we do have bacteria in our stomach, which help us digest food. They allow us to break down some of the food we eat, and we, in turn, provide them with a safe place to eat and live. We're a walking restaurant.

But, and this introduces another form of coevolution, if those organisms in your gut turn out to be not-so-nice, unlike our little friends the gut-bacteria, then we are talking about another form of relationship. One that could be deadly.

Parasites.

One such parasite is the tapeworm. Out of morbid curiosity, I looked to see what the largest tapeworm found in a human being was. I found answers ranging from 28 to 108 feet long. Even eight inches would be too long, in my opinion. Needless to say, they aren't a welcome passenger on our walk through life. They might benefit from the ride they've hitched, but it's certainly of no benefit to us. They are a thief in the kitchen who eats all of our food.

Grossed out yet?

How about antelope dung? Not a living thing, right? So where is the coevolution there? It's not the dung itself, but it involves a certain beetle that likes to lay eggs in the dung. It's called, appropriately enough, the dung beetle. It has a nasty little habit of finding bits of antelope dung, burying it, and laying their eggs in it. It's warm, and when their eggs hatch, it gives the baby beetles something to eat.

Wonderful, right?

Well, one enterprising plant figured out that if its seeds looked like antelope dung, then it could trick the dung beetle into burying its own seeds for it. Of course, it's not a conscious thought on the plant's part, but over many years the seeds evolved. Along the way, any mutation that caused its seeds to look more and more like antelope dung caught the attention of the dung beetle. The beetles helped to select for the adaptations. It reached the point where, not only did the seeds look like antelope dung, they smelled like it too.

We could go on and on with examples, but I think you have the idea. Coevolution can involve relationships that are parasitic, combative, or mutualistic. The end results are never the end. It's constant, like our battle against flu viruses. They evolve to avoid our attempts to squash them, and we tweak our vaccines accordingly.

The flu is still with us, and we do everything we can to stay on top of it. The battle lines haven't moved much.

On July 6, 1917, the Red Baron was wounded during a dogfight by the British pilot Donald Cunnell. The Baron was able to land his plane, but the injuries and subsequent operations kept him grounded for the next three weeks. He recovered and took to the air again. The air battles continued until April 21, 1918. Shortly after 11:00 a.m. on a clear day over Northern France, Manfred von Richthofen met his match. The Red Queen took the Red Baron's hand and led him into history.

Chapter Thirty-Six

Brains

In the 1931 film *Frankenstein*, science created a monster it couldn't destroy. That was the actual tagline of the movie. To give credit where credit is due, it was Victor Frankenstein who created the creature. But before he could reanimate the lump of dead flesh he'd put together, he first had to give his creation a brain. Without it, his monster would have never left the laboratory table nor entered our collective consciousness. Brains are, after all, pretty important.

Science didn't provide you with a brain. Nobody did. Yet it's there. So exactly how did it come to be? What evolutionary forces swept in to form your brain?

It's fair to say that it didn't happen overnight. Before we start wondering how we have the sort of brain we have, let's first look at how it started. What small little collection of synaptic proteins served as the base structure? For this, we can look at the sea sponge.

I bet you never thought you had anything in common with a sea sponge. But you do. We share a common ancestor with them, a very distant common ancestor. The thing with sponges is they do not have a brain or a central nervous system. What they are is a collection of cells. When multicellularity became a thing, the sponges were among the first to give it a shot. For their cells to communicate with one another, they use proteins. It's not a stretch to see how this cellular communication method might have evolved for the cells to band together for protection, using proteins to communicate. Chemical messages pass from cell to cell, binding the entire group together. It's not much different from the way the neurons of your brain communicate with one another.

Imagine an organism, an early organism that consists of billions of cells that can communicate to avoid predation. Or, on the flip side, a group of cells that can direct themselves toward another group of cells with the goal *of* predation. To do so, to move in a certain direction, you must first have a sense of direction. If you're an amorphous blob of cells, who's to say which way is up or down? You're floating along, hoping to bump into something you can assimilate. What if some of these cells become sensitive to light? This collection of cells could communicate to the rest of the colony to move toward or away from a light source. As these cells become more specialized, they remain grouped together. You can think of them as the leaders of the pack. When they want to move, they communicate back to the rest of the colony to move. I say "back" because these cells are now in the front. They dictate the direction in which the colony—we may as well call it an organism now—moves.

Over billions of years, this specialized cell region begins to develop sensory organs. Those light-sensitive cells slowly develop into eyes, and another organ starts to develop. A centralized structure of highly specialized cells that control everything. Then comes backbones—structures to attach muscles to so the organism can move. The early backbones, if you recall from chapter 29, were called notochords. They were basically support structures.

Directing all this was a little group of cells that formed a sort of command center near the photo-sensitive cells. They receive information and pass it along.

As time went by, these organisms became more and more complex. They had to so they could survive. As cells changed and mutated, those changes that were of a benefit persisted. It was trial and error on a grand scale. One of the most important, if not *the* most important adaptations, was this little group of highly specialized cells. As the organism grew more complex, there was a lot more for these cells to do. Their jobs demanded more of them, especially when the organism moved away from the environment and it spent its first few billion years on dry land.

These cells were responsible for regulating breathing, basic motor skills, and controlling the heart.

If nature was Dr. Frankenstein, then this specialized cell region was an early beta version of our brain. A 570-million-year-old beta version. We now refer to it as the hindbrain. It's still there, buried beneath everything else that has developed to become your brain today. It's responsible for many of your basic motor behaviors and movements. Just like it was 570 million years ago. Your medulla, pons, and cerebellum are found there.

But of course, nature isn't content to leave things alone. It must continue to tinker. Especially when it comes to keeping the organism competitive. Around 250 million years after the development of the hindbrain, *more* specialized structures appeared. These were built upon the preexisting hindbrain to deal with the fact that it's a dog-eat-dog world out there. This region is now called the limbic region. It is responsible for more complex behaviors, such as fighting, emotions, and behaviors that promote reproduction. Reproduction is, after all, what it's all about. Your amygdala is a part of this region, as well as your hippocampus.

Nature wasn't done yet. Now that organisms could take care of their basic survival needs, it was time to give them something to think about.

It was time to create your neocortex.

I'm going to leave Dr. Frankenstein behind now and move on to another analogy. Think of a city and how it develops. You start with a few dirt roads and perhaps some simple shelters. These represent your hindbrain. Your basic needs are met here. You have shelter and a direction. The dirt road tells you where to go and how to get home.

Over time, as more and more people find their way to your little town, other structures begin to emerge to suit the town's needs. A little store pops up, a barbershop, and eventually a saloon and sheriff's office to keep everything under control.

You can call these additions to the town the limbic region. There's a lot more activity, conflict, and conflict resolution going on now. As more people move in, the roads are expanded, and new structures emerge. The new structures are tall and dominate the sky. Skyscrapers appear almost overnight as well as banks, insurance companies, art galleries, and air control towers.

This is your neocortex, the most specialized area of all. It took a long time to get here, but now that it is, there's no going back. If you look closely, everything that came before is still there. You can still see those first dirt roads, paved over now, and the barbershop and saloon are still present yet under different management. Even the sheriff's office survived, but now a large federal building sits above it. What was once one floor is now ten.

Without your neocortex, you wouldn't be able to talk, paint, think, or do any of the higher cognitive things you can do. You'd still be scurrying around the trees, hoping to avoid predators. Now you're able to sit twenty floors above it all, twenty floors that you created and designed. We can all raise a toast. We've come a long way.

Now that we have some idea about the order of things, how nature built your brain over time, there's a question to be asked. How did it expand so quickly? Granted, a few million years doesn't sound fast when you say it out loud. But when you consider your hindbrain first appeared five hundred million years ago, your neocortex underwent a rapid expansion two to three million years ago. Much faster than the mammals you share the planet with.

Let's think about it for a moment. Something must have accounted for it. Something in the environment presented enough challenges to shape it into the formidable force it is today, evolutionarily speaking, of course. It's not like we have reached the apex of brain development. Nor does it mean it couldn't have happened in an entirely different way. We could have spent the last two to three million years with the smaller brains we started with.

You have to remember, we shared this planet with the dinosaurs at one time. Well, when I say *we*, I mean our mammalian ancestors did. They scurried about beneath tree limbs and stuck to the underbrush to keep out of sight. Those were challenging times for sure. Our ancestors had to be keen, and they needed to be cunning. They also required well-developed senses. Larger brains would help with that. As our eyesight got better, so too would our neocortex.

But then, as luck would have it, the dinosaurs died, and the mammals inherited the earth. That was roughly sixty-five million years ago. After that, it was time for mammals to rise up and take their place at the head of the table.

Flash forward some sixty million years, and that is when we last shared a common ancestor with the chimpanzee. Five to six million years ago, their ancestor went one way, and ours went the other. The homo genus continued to evolve, travel to Europe, and spread. We didn't do this alone; we did it together in groups. Group dynamics take a lot of energy. You have to remember things. You have to communicate. You have to face life together as a cohesive unit. This all takes computing power—quite a lot of it.

Another possible contributing factor is fire.

In our early days, our ancestors ate raw foods. Their diets consisted of nuts, plants, and raw meat. One needs energy to process that food. A lot of it to break it all down into something the body can use. Have you ever had a huge meal and felt tired and lethargic afterward? That's because your body consumed energy to deal with the food you swallowed. Feed it raw food, and it needs more energy.

Along comes fire. Not only could it be used as a heat source to deal with the cold European nights, but it also enabled us to cook.

We can only guess how the first cooked meal was discovered, and it was most likely an accident. Our culinary appetites were unlocked once we got a taste for it, and we never looked back. The great thing about cooked food is it has been broken down for us before it ever gets to our stomachs. The discovery of fire may have also coincided with our use of tools to cut animals up. Cooking meat unlocks nutrients.

Fire may have played into our development of larger brains. Fire, and the cooking of food, happened well before modern humans ever walked the earth. It started as a flicker in a distant, pre-*Homo sapiens* past.

And two hundred thousand years ago, we modern humans stepped into the spotlight, ready to take on the world. And to think, about seven million years ago, the human brain was once a third of the size it is now.

You can't judge a book by its cover. Nor can you judge an animal's intelligence by the size of its brain. Call it brain discrimination. You see a small animal with a small brain, and it's obvious you're not dealing with a rocket scientist. Douglas Adams handled this brilliantly in *A Hitchhiker's Guide to the Galaxy*. In it, mice were the brilliant ones. The earth was built by them as a giant computer to answer the meaning of life. The joke was on us. Silly apes. Adams could just have easily used crows. Crow brains are different than ours. Our last common ancestor was roughly three hundred million years ago. Since that time, we've evolved the prefrontal cortex, while crows evolved an area called the nidopallium caudolaterale (NCL). Crows have shown remarkable abilities when it comes to reasoning, even abstract reasoning, both in nature and in the lab. They can use tools and recognize faces. They even have the ability to deceive.

It's not about size. It's about structure.

Studies have shown that our brains, modern human brains, have shrunk slightly in size. Some estimates put the area we've lost over the last twenty thousand years to that of a tennis ball. There are several theories as to why. Perhaps it's because we've domesticated ourselves. The survival pressures we once faced are no longer there. Are we getting dumber? Well, that's debatable, but it could be the brain is becoming more specialized. Or it could be we are outsourcing a lot of the brain's functions just as we did with our stomachs when we discovered how to cook.

Outsourcing our brains sounds preposterous, right? It's not, when you consider the role computers and the internet now play in our lives. Artificial intelligence is not too far away.

Reenter Dr. Frankenstein.

Chapter Thirty-Seven

Convergence

The world needs heroes. Kids needed heroes during World War II and the Cold War. Like Captain America and Superman. Noble heroes with noble goals. Both heroes came from different universes. The universes of DC comics and Marvel comics. The pages of the comics produced by both publishers were filled with more characters than one could count. The superheroes within were as amazing as they were colorful. There were characters to fill every niche, be it the swamp, ocean, or sky. If your character lived under the sea like Aquaman, there were certain things he had to be able to do. Most importantly, he needed the ability to breathe underwater. Some characters could fly. There were many similarities between the heroes that stalked the pages of DC and Marvel. Think of these two separate publishers as two distinct environments, complete with their own challenges, villains, and conflicts.

Here, in the real world, it's no different. Take an environment, any environment or ecological niche, find a similar one on the other side of the world, and you will see similar creatures. These creatures evolved separately, but they share structures that perform the same functions. It's so prevalent there's a term for it.

It's called convergent evolution.

Every living thing on this planet shares a common ancestor at some point in the far distant past. In some cases, that common ancestor existed millions of years ago. Over that stretch of time, you will see similar adaptations that were not present in that common ancestor develop separately.

A classic example of this is the ability to fly. Both birds and bats fly. Their last common ancestor could not. Over the past few million years, they separately developed the ability to lift themselves off the ground to hunt and escape being hunted. They can both do this, but the way they do it is different. The underlying mechanism that allows them to fly is built differently but uses the same bones present in their last common ancestor. They both have arms with the same bone structure they inherited from that ancestor. It is practically universal among vertebrates. You and I share it as well. Our forearms are made up of an ulna and radius. It's the same with birds and bats. When you get to their fingers, their metacarpals, and phalanges, things are much different. For us humans, our metacarpals and phalanges evolved to allow us to grasp things or pick items up. In birds, they developed into a different structure, more like an extension of the forearm to which their feathers are attached. Bats have incredibly long fingers. The bones their leathery wings are attached to are actually their metacarpals and phalanges. The same ones we have only longer.

This underlying bone structure, because it is the same, is considered homologous. We all share this structure because it was present in our common ancestor. The wings themselves, those structures that allow for flight in both bats and birds, are different. They are considered analogous structures. They evolved to perform the same function separately in different species. Such structures evolved in unrelated species, which serve as examples of convergent evolution.

Let's go back to the echolocation we discussed in an earlier chapter. Bats and dolphins do this by emitting short bursts of sounds that bounce off whatever objects or obstacles happen to be in front of them. They can judge, unconsciously, how large and how far away the object is by the intensity of the returning sound wave. People can do it too. By using a series of clicks, a person, when properly trained, can navigate the world, even ride a bike by using a form of echolocation. While this is a skill that you or I could learn given time, in the case of bats and dolphins, it is an example of convergent evolution. They may both be mammals, but the most recent common ancestor between the two lived around sixty million years ago and did not have this ability. It evolved separately. Because bats and dolphins are built differently from one another, their use of biosonar is slightly different.

Then there's the human eye. It's intricate, complicated—and flawed.

In the chapter on mistakes, I mentioned you have a blind spot. You might not be aware of it because your eye movements compensate for it, and your brain fills in the gap. But it's there. It's there because your optic nerve evolved to plug into the eye in front of the retina. The retina is the layer of light-sensitive cells that coats the inner part of your eye.

According to some estimates, the evolution of the eye happened over forty different times in unrelated species. By unrelated, I mean they are separated by millions of years and evolved into entirely different species. In cephalopods, like the squid, the eye is wired differently. It is wired much better than our own, with the optic nerve cells coming in behind the retina. So no blind spot.

In some cases, convergent evolution is obvious, as in the case with bats and dolphins—two separate species that have evolved a similar biosonar method to navigate the world. And, like the evolution of eyes, these are all pretty clear-cut examples. It's not always that plain. In some instances, it could be a case of re-evolution. This means certain genes that had been present in a common ancestor were somehow switched off. This could happen for any number of reasons, the most common one is some sort of mutation. Long after that ancestor evolved into two separate species, say millions of years later, another mutation comes along and turns the gene back on. What might seem to have been a case of convergent evolution might simply have been a faulty switch.

In his book *Wonderful Life*, biologist Stephen Jay Gould famously said that if you rewound the tape of life and started it again, you might see an entirely different story play out. The world, as we know it, might be populated by different designs and different solutions. We could get into a whole philosophical argument here about determinism and theories about time and reality, but let's stick with biology. I can see why Professor Gould would say this. If the evolution of life started over, who's to say the mutations that shaped the eye would happen again? Nature might find another way to use those early light-sensitive cells. Who's to say legs would have evolved? Or noses? Any number of things may have occurred differently.

Convergent evolution plays a significant role in the history of life both on this planet and every other life-sustaining planet out there. And if there is another planet like this one out there, it might be inhabited by creatures very much like us—creatures with eyes, mouths, ears, and two legs. Or, then again, that life might be nothing like anything we have ever seen or can even imagine.

Chapter Thirty-Eight

Genetic Drift

I want to tell you about a little town that sits in the shadows of an extremely large mountain. It's not a real place. But if it were, it would be known for its frogs, and because of this, we'll call it Anura (which just so happens to be the taxonomic rank of frogs). There are two kinds of frogs in Anura, some with blue pigments and others with green. Other than their color difference, they are almost exactly alike. For some reason, Anura is the only place on earth this particular species of frog exists. There are no people, it's quiet, there's plenty of food, and the other animals pretty much leave them alone. Once in a while, a wolf will come by, and when it does, it has an eye for the blue frogs. No one knows why, and no one has ever studied this, but it does seem to be the case and, as a result, the population of green frogs is slightly higher than that of the blue frogs. 60 percent of the frogs in Anura are green, and 40 percent are blue. It fluctuates somewhat, but those pesky wolves keep the percentages fairly consistent. You could say that natural selection favors green frogs. If an entire pack of wolves were to come by, it could mean bad news for Anura blue frogs.

On this particular day, there's a storm brewing—a big one. The light of the moon has been blocked out by gigantic storm clouds.

If you were a frog who appreciated light displays, you would find the lightning spellbinding. Until it strikes a tree, of course. And on this particular night, that's precisely what happens. Not only does it shatter the tree, but it starts a fire. Anura has suffered a dry spell recently. As a result, the fire spreads.

Anura has never seen anything like it. The fire leaps from tree to tree. If you were to pass overhead, it would appear as if the entire area was engulfed in a sea of flames.

The next morning the creatures of Anura slowly come out to assess the damage. There are quite a few charred trees, and large areas of the forest appear to have been dipped in black paint.

Life in Anura resumes. But one thing has changed. There are now more blue frogs than green. Most of the green frogs died in the fire. If one were to count them, the ratio would now be 80 to 20. Out of every 100 frogs, 80 of them are blue.

If you were a biologist studying Anura's fauna after the Great Fire, you would want to know why there were more blue frogs than green. You would assume that the blue frogs had the advantage and that natural selection was doing its job.

But that's not the case, is it? The truth of the matter is that natural selection has nothing to do with the predominance of blue frogs. A chance lightning storm is responsible. A fire wiped out most of the green frogs. Not because they were slower or weaker. It was simply because they were caught in the fire and couldn't escape.

This random, dare I call it "selection," process is known as genetic drift.

Natural selection plays favorites. As animals change and pass on their adaptations, populations grow. You can look at a population and theorize why particular traits have survived by looking at the environment they've survived in. Genetic drift doesn't play favorites. It's blind. It moves in, randomly stumbles around, and then leaves.

Look at it like this. Let's say I were to pour fifty blue marbles and fifty green marbles into a jar, shake it up, and set it in front of you. I then tell you to close your eyes and remove fifty marbles from the jar. When you're finished, we count what's left in the jar and see it contains 33 green marbles and 17 blue marbles.

That's how genetic drift works on small populations. Suppose I had asked you to remove fifty marbles *without* closing your eyes and you removed more blue than green. In that case, that could be likened to a different type of selection. There was something about the blue marbles that caught your attention, like a peacock's feathers. If you tried to pick out marbles and the blue marbles were coated in oil and slipped out of your fingers, that could be attributed to natural selection. In this case, they had a trait, a slippery, oily surface that allowed them to survive.

But genetic drift is random. The green frogs, who previously had an advantage when wolves came by, were wiped out not by selection but by chance.

Let's get a little technical for a moment. We already know that natural selection likes to work with genes. If a gene gives you longer legs and those legs allow you to live longer than your neighbors, long enough to reproduce, then you pass those genes on. Your offspring will be more likely to have longer legs. What caused the long legs is usually a mutation. We have two sets of each gene. These two alternative forms of the same gene are called alleles. If a population of animals has legs of the same length, that can be traced back to alleles that account for leg length. If an organism has two leg length alleles that are exactly the same, then, by right, everything stays the same. If one of the alleles is different, there's a variation there. That variation in the leg length allele may result in a visible difference. Such as longer legs.

This is what is measured by genetic drift—changes in allele frequency. Natural selection picks those alleles that allow for an organism to survive. Genetic drift is entirely random.

Let's go back to our colony of frogs in Anura. If you recall, the wolves liked to eat the blue frogs when they came by. Natural selection would dictate that, over time, that population might only contain green frogs. The blue frogs would become extinct. But along came a lightning storm and forest fire, which wiped out most of the green frogs. Measure the allele frequency after that event. You'll see a greater percentage of the allele that accounts for blue frogs in that population. The genes have drifted to that of blue frogs.

The Great Anura Fire left a population of blue frogs in its wake. Step out of Anura, and you might find that the rest of the world contained naturally selected green frogs.

If there is one thing we can always count on, it's change.

Part Four

The Wonder of It All

Chapter Thirty-Nine

Self-Directed Evolution

Throughout this book, we've been focused on the past. What brought us here, what we know about it, and how we know it. We've talked about pre-Darwinian thinkers, Charles Darwin himself, and Alfred Russel Wallace. We've looked at our humble beginnings as prokaryotic cells that clumped together to form eukaryotic cells and eventually multicellular organisms.

What we haven't yet looked at is the future. We are standing on the stepping stone of an uncertain path: a path both exciting and a little scary.

In 2010 a team led by biochemist Craig Venter created the first synthetic cell. When I say "synthetic," I don't mean they constructed it from scratch. What they did do was no less remarkable. They created a self-replicating cell that had never existed before. It started with the genome of a goat pathogen and the empty cytoplasm of a bacterial cell. Venter's team successfully injected the genome into the bacterium and waited.

It began to replicate. Over and over again.

They also added a little something else to the DNA they constructed—a watermark. More specifically, they added a series of watermarks. Why a watermark? To identify descendant cells should the cells organize a prison break.

Of course, Venter and his team have ensured these created cells remain in the lab. You won't encounter them. Not these cells. But the next time you get a flu shot, that vaccine, the dead cells that make it up, could be developed by the same process.

It's more than a little mind-blowing.

I said that we were going to look forward to what this may mean for our future, but first, let's have a quick recap of what has led us to this point.

Almost 14 billion years ago, something happened. Something that has been attributed to a "Big Bang." Others have called it the great expansion. It's most likely a little of both. Something caused a hot dense little ball of matter to expand rapidly in the blink of an eye. Ten billion years later, as things cooled down, our little solar system was formed. This formation included our little blue sphere. The one we call home. Toss in another billion years, and life appeared and began to populate our early oceans.

Life left those oceans around 400 million years ago. Creatures of all shapes and sizes struggled, survived, and evolved. Modern humans appeared on the scene 200,000 years ago. From the Big Bang to the human brain in slightly under 14 billion years. We haven't been here very long. But the things we've accomplished in this short period of time, using our brains, is nothing short of miraculous. My favorite example of the power contained within our brains starts with the Wright Brothers' first powered flight in 1903. They spent only a few moments in the air. Sixty-six years later, we were walking on the moon.

In 1876 Alexander Graham Bell and Thomas Watson created the very first telephone. In 1878 the Bell Telephone Company was born. Eleven years later, a Kansas undertaker named Almon Strowger added a rotary dial to Edison's device. In the 1960s, the first push-button phones hit the market.

Why am I talking about phones in a book about evolution? To prove a point. Everything evolves. Cars, buildings, bridges, and phones. Take every phone design, all of its permutations, and tape them to a wall. You'll need a big wall. Step back and look. There's a story told there. You can see the changes from Alexander Graham Bell's first liquid transmitter to the iPhone. And today, our phones even talk back. There no longer needs to be a person at the other end.

My point in all this is we've come a long way technologically in a very short period. Biological evolution has brought us far. It's been a trip that has lasted three and a half billion years.

Evolution by natural selection is a slow process. A painfully slow process. What we are now doing in the lab, and to our bodies, is pushing it along much faster in leaps and bounds. Centuries ago, we made our faulty vision better by creating eyeglasses. Decades ago, we created hearing aids. Today we can fashion mechanical appendages to replace lost limbs. In some cases, these new arms and legs are superior to the ones we've lost. Next will come artificial eyes.

I recently read E. O. Wilson's book *The Meaning of Human Existence*, published in 2014. Professor Wilson calls this stage of our evolution "Volition Selection" as opposed to natural selection. We are reaching the point where we will now be able to dictate what biological features, and nonbiological enhancements, accompany us into the future.

Genetic defects may become a thing of the past. Cystic fibrosis, muscular dystrophy, and breast cancer may one day be eliminated. The more we understand our genetic code, the more we will tinker with it. We simply cannot help ourselves. We are tinkerers, and we will continue to do so until something goes wrong. History hasn't exactly shown us to be the most responsible creatures when it comes to new technology. We often test it out on the fields of war first or look for ways to increase the size of our wallets.

Mankind has only just begun to unlock the secrets hidden within our DNA. As we move from gene to gene, we will start to see how it all ties together and where evolution made a few mistakes. It will be within our power to correct those mistakes. Will it be at the cost of our humanity? That remains to be seen. But we are curious creatures, and if there is a large stone to turn over, you can be sure that someone will come along with a tool to turn it over sooner or later. It's what we find beneath it that may be cause for concern.

I'm optimistic about our future. I have little doubt that we will eventually turn our DNA inside out and find a way to increase the length, and hopefully the quality, of our lives. Science will accomplish this. Understanding how we are put together will enable us to construct new things similar to ourselves but made out of something other than flesh and bone. Artificial intelligence is peeking back at us just over the horizon. We could be designing the next branch of human evolution, a companion to accompany us into the future.

Chapter Forty

Are We Unique?

The human body is a machine which winds its own springs.

—Julien Offray de La Mettrie

In 1748 a French philosopher named Julian Offray de La Mettrie wrote a little book called *L'homme Machine*, which in English is *Machine Man.* We learned way back in chapter 6 that La Mettrie's *L'homme Machine* was classified a subversive book because of its controversial stance on free will. What La Mettrie wished to convey was that mankind is no different than anything else. The "anything else" is trees, flowers, birds, and rocks. We are only different in that we are put together differently. We have a circulatory system and are mobile. But other than that, we're not so special. In essence, we are just biological automatons. We go about our daily business, living, breathing, eating, because it is what we are designed to do. And he wasn't saying that our design was the result of planning. Nature did it in all of her blind, random glory. Not only that, but she pulled another fast one on us. While we might think we have conscious thoughts, La Mettrie was happy to tell us we do not. Consciousness, or what we think of as consciousness, is a byproduct of the organic changes going on in the folds of our brains. It's an epiphenomenon. We are no more in control of those apparent thought processes than we are a storm on the horizon.

He paints a gloomy picture. A century before La Mettrie felt impelled to knock us off our evolutionary pedestal, René Descartes told us that we were special. According to Descartes, "*animals are mere machines but man stands alone.*" We were different *because* we had brains, a consciousness, and we could perform actions on our own. We were separated from animals by divine right *and* design.

The thing is, we *feel* like we're special. The things we have overcome as a species, the obstacles we have conquered, speak to our unique place in nature.

Or so we like to think.

So in what ways are we special or unique? Is it because we can think, as Descartes said? Or is thinking just a chemical process that directs our actions as La Mettrie would have us believe? And what does it even mean to think or to be conscious? You might be shocked to know we don't really understand how it works—any of it. Science hasn't been able to touch it. We are making slow progress. Until we can declare either Descartes or La Mettrie the winner of this debate, we can all rest assured our ability to reason, and our self-awareness, appear to be ours and ours alone.

Or is it?

In 1838 Charles Darwin visited the London Zoo to watch the orangutans. There were two orangutans present on this day, and they were given a mirror. To the astonishment and pleasure of those looking on, nobody more than Darwin, the orangutans were mesmerized by their new toy. Darwin described the encounter with much amusement in some notes:

They at last put a hand behind glass at various distances, looked over it, rubbed front of glass, made faces at it—examined whole glass—put body in all kinds of positions when approaching glass to examine it.

Looking at oneself in the mirror is something we have always pointed to as uniquely human. Even infants can discern that the baby looking back from the borders of a mirror might be themselves. It's the perfect test of self-awareness. It's called the mirror test.

The mirror test, or mirror self-recognition test, was developed by psychologist Gordon Gallup Jr. in 1970. Gallup was trying to prove self-awareness isn't unique to humanity. Recognizing oneself as an individual separate from the environment is something other animals can do.

Just as Charles Darwin witnessed in 1838 with the orangutans. They, just like us, know that the stranger in the mirror isn't a stranger at all.

Our closest relatives, the chimpanzees, like to play with their reflection. They seem to recognize that the image they see in the mirror is equivalent to their own body.

They are not the only other animal who has passed the mirror self-recognition test. Also on the list are dolphins, magpies, and Asian elephants.

So the hunt for intelligent life need not look toward the stars but right here on earth. We are not alone.

Again, let's wind the clock back a little over fourteen decades to a time when Darwin's *The Descent of Man* hit the bookshelves of London. Eager readers opened the covers to find that not only were they *not* special when it came to being a smart animal, and that any difference is only a little more than skin deep:

There is no fundamental difference between man and the higher mammals in their mental faculties," and all the differences are differences "of degree, not of kind.

Shortly after publishing *The Descent of Man* in 1871, Darwin published another book, this one called *The Expression of the Emotions in Man and Animals*. In it he outlined, categorized, and listed all of the emotions we share with other animals on the planet. All of which share a common ancestor with us, without exception.

The very first electric razor was invented and patented in 1928 by Jacob Schick. Razors have come a long way. Perhaps as far back as the sixth century BCE if we are to believe the Roman historian Livy. Before the razor, we may have found facial hair grooming challenging, but we weren't strangers to tools. Tool use was once considered to be a uniquely human activity. We can again thank our relatives, the chimpanzees, for dispelling this myth.

Primatologist and anthropologist Jane Goodall has been studying chimpanzees since the 1960s. Before she came along, it was thought chimpanzees were vegetarians. She discovered not only was this not true, but they used tools to hunt for meat. Even if the meat they searched for was termites.

Writing in the mid-eighteenth century, the philosopher Jean-Jacques Rousseau put forth the preposition that man was good by nature but had been spoiled by civilization. This innate "goodness" makes itself evident in small actions. Those actions inevitably reveal hidden truths. We hate to see others suffer. Studies have shown that rats will save a drowning friend. In an interview conducted by Andrea Sachs for *Time* magazine, Jane Goodall talked about one of the first chimpanzees to lose the fear he initially felt by her presence. Chimps are very wary around humans at first. This chimp, David Greybeard, was pushing through the forest, and Jane was following him. It became rather thick, and she was afraid she was going to lose him. According to Goodall, she found Greybeard sitting and waiting for her. When she saw him, she picked up a nut and held it out for him. Instead of accepting the nut, he dropped it and gently squeezed her hand in a sign of reassurance. She was not lost; he knew the way.

That moment when Jane Goodall looked into the eyes of David Greybeard is reminiscent of the moment when Charles Darwin looked into the eyes of the orangutan at the London Zoo almost a century and a half earlier. There was a moment of recognition there. Recognition that we are not alone on this planet. David Greybeard's act of reassurance displays an awareness of another's mental state. This requires conscious knowledge.

Perhaps we are not so unique after all.

Chapter Forty-One

Are We Still Evolving?

These days, a common claim is that evolution by natural selection, at least when it applies to you and me, is no longer that big a deal. The argument is we've pushed natural selection aside and taken the reigns of our development. Through technology, medical advances, and our different cultures, the things that once drove our evolution are gone.

Charles Darwin's theory is simple in its elegance. Nature selects for those adaptations that allow an organism to survive long enough to reproduce. Those adaptations are passed on and, over time, dominate the population. It can be an isolated population like one that lives on an island. Darwin saw this with the finches and tortoises in the Galapagos. Take a species, split it into two groups, separate them by an impassable mountain range, and step away for a few million years. Peek in on them, and you may find so many differences between the two that they are no longer able to mate with one another. Give it a few more million years, and they may not even look as if they ever shared a common ancestor. The next time you pet a cat, know that your shared common ancestor once walked the earth an estimated one hundred million years ago.

Natural selection is subtle, and it doesn't perform magic tricks. It won't pull a rabbit out of a hat, but it will turn the simple celled organism living in the hat into something else if you give it enough time.

Now back to the argument that natural selection no longer plays an essential part in our development as a species. It's very premise says we've somehow stepped out of nature and are now immune to its effects.

Nature doesn't mean the jungle or a less industrialized culture. If you hop into a space shuttle, one that can support life for a few million years, you're still affected by the forces of nature. You will still adapt to your environment. Selective forces will always be at play. You can't get away from it. And there is no set speed. The road of evolution doesn't care about speed or destination. It doesn't care about anything at all. Our job, as a species, is to stay on the road. Or die trying.

Malaria is a nasty disease and one we've been battling for centuries. Long story short, there's a little protozoan who loves to hitch a ride on mosquitoes unannounced. If an organism has this protozoan in its blood, and a mosquito comes to draw blood, it will leave with the protozoan. The protozoan sets up shop in the mosquito's gut and breeds. The next time the mosquito lands, there is a transfer of blood. If that transfer of blood involves you or me, that protozoan heads to the liver where it lives, reproduces, and eventually bursts into the blood to attack the red blood cells.

That's when the trouble starts.

While we as a species are throwing everything we have at it to stop it, nature has been doing so as well.

You may be surprised to know some populations in Africa have benefitted from adaptations that have provided them with a means to combat this disease. One of those adaptations we can point to is sickle cell anemia. While it's not something you want to have under normal conditions, when it comes to malaria, it's extremely effective against the malaria parasite. Those red blood cells are no longer disc-shaped; they are sickle-shaped. Think of a crescent moon, and you'll have the right idea. The parasite, when it encounters a sickle-shaped cell, can't do anything with it. The red blood, while handicapped from performing its normal duties, is immune to the parasite's attack.

In populations where malaria runs rampant, those individuals with the sickle cell gene have an advantage.

Once it happens, if it happens *and* provides an advantage, nature has something to select for. And then that gene, or its mutation, might survive.

If you can drink milk without feeling gaseous or eat a bowl of ice cream without experiencing cramps, you might be a mutant. Those of us who can are the exception. Seventy-five percent of the world's population can't do what we do.

It's all because of four little genes. Or alleles if you want to be more specific.

When we are young, we produce an enzyme called lactase. This enzyme enables us to break down lactose in milk. Lactose is a carbohydrate, and without lactase to break it down, we have a tough time digesting it. Because most mammals are introduced to the world with their mother's milk as their chief form of nutrition, it makes sense. We produce lactase to break down the lactose. As we get older, lactase production is stopped.

For some of us, that is. For others, lactase continues to be produced. It allows us to enjoy Friday night pizza, milk in our cereal, and big bowls of Ben & Jerry's Peanut Butter Fudge.

This ability to generate lactase well after we have been weaned off of our mother's milk is due to a mutation in our genes, which occurred around 7,500 years ago. It started in Europe with cattle herders. Milk remained a staple of their diets, and so they evolved to accommodate it. Thus, the four genes I mentioned mutated to allow for lactase to be produced into adulthood. To show that nature often solves the same problem the same way, this ability to produce lactase while the rest of the world's population shuts it off has been found in other cultures. You can find these mutations in Tanzania, Kenya, and Sudan. And today, if you're in the United States, the numbers are flipped. Instead of 25 percent of us who produce lactase as an adult, it's now 75 percent.

Since we're on the subject of digestive enzymes, there's another example of human evolution in action. This involves the breaking down of starches found in carbohydrates. Populations that consume a lot of carbs also produce an enzyme called amylase. And amylase production is tied to a gene called AMY1. If your culture consumes more protein than carbs, you'll find the AMY1 gene less prevalent. Again, nature selects for what works when presented with the option.

And if you're part of the Inuit people of Greenland, you also eat a lot of fatty acids. And the digestion of fatty acids is linked to mutated alleles present in Inuit ancestry.

You know the old saying "you are what you eat." You literally are *because* of what you eat.

I'll give you one more example, an even more recent one.

In Tibet, especially the higher elevations, the air is extremely thin. So thin, in fact, that you and I would have a very hard time breathing. We certainly wouldn't go out for a morning run. If I asked you how you could fix it given a chance, you might tell me to either provide everyone with oxygen tanks or to change their biology so they enjoy a higher concentration of oxygen in the bloodstream.

Therein lies the predictive power of evolution. Researchers have found that the people in Tibet have evolved to allow for a greater concentration of oxygen in their blood. Just as you would predict. It's more evidence of natural selection working on a human population placed in a harsh environment. Don't forget that natural selection will choose an adaptation or set of traits that allow an organism to live long enough to produce copies of itself. In this case, a mutation that will enable oxygen-rich blood for the people of Tibet, at least above normal oxygen-rich blood, gives a child a better chance of surviving into adulthood. So that allele, the one that accounts for a higher concentration of oxygen, persists and propagates.

As a species, we have done a fantastic job of reducing the stress in our lives and on our bodies. We have our large brains to thank for that. Even if they are shrinking a bit, it's another sign of evolution in action, even though we are not quite sure why this is at the moment. Our ability to mitigate stress has done amazing things for us as a culture and an organism. Whereas certain diseases would have wiped out large groups of the population in the past, we can now combat these diseases before they have a chance to do so. We develop cures and solutions before natural selection has a chance to do so for us. And, it's safe to say, had we not developed the cures, natural selection may never have had a chance to save us. We could have gone the way of the dinosaurs a long time ago. We might still if an asteroid barrels into us. But barring an asteroid collision, we've managed to do just fine without having to wait for a random mutation here and there to save us.

This doesn't mean we are not evolving or that natural selection has nothing to do. The way I look at it, and I would argue the way it should be looked at, is that every advance we make, everything we do to extend human lives, is natural selection at work. It doesn't have to be tied to a mutation in our genes. It can also be attributed to the amazing things we create and accomplish. We are still the products of nature, working in nature, and manipulating nature to our advantage. There are selective forces at work. They follow the laws of nature and allow us to survive a little bit longer than our ancestors were able to. From this humble little planet, we may eventually spread out to populate more humble little planets. On those planets, we will encounter new challenges and chances to evolve.

Chapter Forty-Two

Just a Theory

A few years ago, I was at my son's football practice and had a book about evolution on the grass next to my chair. One of the other parents walked by, looked down, saw the book, and frowned. She said, "you should read the Bible instead," and continued on her way.

It wasn't an isolated incident. Last year I was at a dinner party. During the party, I happened to mention my podcast about evolution. One of the guests rolled her eyes.

"That's ridiculous," she said. She jumped right in with guns blazing. Up to this point, the dinner party had been a pleasant one.

"What is?" I asked.

She rolled her eyes again. She was very good at that. "*I* didn't come from a monkey."

I wanted to tell her she was right. She didn't come from a monkey. Monkeys, chimpanzees, gorillas, and everyone sitting at that table had descended from a common ancestor—some long-ago primate who enjoyed the sun in Africa instead of the seasons in New Hampshire. I wanted to tell her around 25 million years ago this common ancestor split off into groups. Each group walked into different environments and faced different situations to which they had to adapt. Over that 25 million years, natural selection shaped one branch into chimpanzees and another branch into humans. There were other branches too. For all we know, there have been too many to count. Some branches gave birth to gorillas, rhesus monkeys, and orangutans. And don't forget the

Neanderthals whom we coexisted with for quite a long time. The last of them apparently died out around 40,000 years ago. Neanderthals were not an intermediary between our common ancestor and us. They were another branch that coexisted and shared Europe with us for an estimated 5,000 years. If things had been different, we could have been the ones who died out, and a bunch of Neanderthals would be sitting around this dinner table instead.

I could have explained it all to her, but I didn't. All I said was, "You should listen to the show."

Her response, accompanied by yet another eye roll as she reached for the bread, was, "It's only a theory."

Only a theory. How many times have we heard that over the years? Is she right? We often hear it referred to as the "theory of evolution by natural selection." It's never presented as the "law" of evolution by natural selection. So what gives? I have theories, too. I think someone sneaks into the house late at night and moves my keys so I can't find them in the morning. I can call it a theory and tell everyone I know about it, much to their annoyance. The same goes for someone who has theories about life after death. There's no way to prove or disprove their theory, yet they are justified in calling it a theory. If that's the case, then isn't calling Darwin's great idea a "theory" somewhat vanilla? Isn't it on no better footing than my theory about why I can never find my keys no matter how many locks I put on my door to keep the "key-moving" perpetrator out?

Well, you'll be glad to know that Darwin's theory is on much better footing. It's on solid ground.

Let's start with a definition. The Merriam-Webster online dictionary defines a theory as "a plausible or scientifically acceptable general principle or body of principles offered to explain phenomena."

Fair enough, right? At the beginning of the sixth century, Copernicus began work on what is now called his Heliocentric Theory. His theory is that the world, our world, and all other planets revolve around the sun. Before this, everyone thought everything revolved around the earth. Including the sun. We don't refer to this as the Heliocentric Law. It's a theory. A scientific theory. Before that, it was a hypothesis. Copernicus looked up, began to think about the sun, the moon, and the planets, and started to make calculations. He thought, and heretically so, that maybe everyone had it wrong and the earth wasn't so special. Perhaps our place in the universe wasn't the center of everything. Maybe we were just one more chunk of rock revolving around the sun. For the past five centuries, science has tested this theory and supported its conclusions. There is very little chance of it ever being proven wrong. It went from a simple hypothesis, was subjected to many different tests, and stood up to them all without exception.

This is what science does when it tests Darwin's theory of evolution by natural selection. It's not a simple hypothesis anymore. It was once. There was a time when Darwin had an idea he wished to pursue. It's why he spent years ruining his eyes in the study of barnacles and barely perceptible differences in shape and color. Those who study his theory today have mountains of evidence to sort through from the fossil record to our DNA. It's how we know the plight of the Neanderthals might be that we assimilated them. There is Neanderthal DNA in our genes. A simple DNA test can confirm just how much. If you're of European descent, it might be quite a lot.

Organisms evolve and change. That is a fact. We see it every day, both in the lab and in nature. It's why professional breeders are so successful. They know what features to look for in a prize dog and which features to discard. They understand heredity and change. That's evolution. It might be artificial selection, but it achieves the same result. There can be no denying it is a fact. If it wasn't, you would never have the pleasure of watching the Westminster Kennel Club Dog Show. Darwin's ideas around natural selection, and not artificial selection, as the engine behind evolution in nature, is the theory. It started as a hypothesis when he first set foot on the Galapagos Islands and observed the differences in bird and plant species on different islands. After years of testing and evidence to support it, that hypothesis became a theory of explanation.

Evolution is a fact. It's not just sort of true; it *is* true. We see it every year with each new strain of flu virus. It is why you need a new flu shot every year. It's not because a new flu virus was suddenly created out of whole cloth and unleashed on the world. A new strain of the virus doesn't appear out of thin air. That's what a new strain is. It's a previous virus strain that has evolved. Our immune systems will not recognize it or know how to handle it when it's encountered. That's evolution.

We have the fact of evolution, and a little over 150 years ago, Charles Darwin had a hypothesis as to how it all worked. He didn't quite buy the hypothesis that Jean Baptiste Lamarck put in front of the scientific committee. He suggested that animals evolve because they pass traits on to their offspring which they formed during life. If that were the case, a weightlifter would pass on the trait for large muscles to his child. Lamarck's hypothesis tried to explain why giraffes have long necks by pointing out how they stretched and strained to reach the next level of leaves in trees. Darwin couldn't see how any of that was possible. Not when he stepped onto 4 of the 16 Galapagos islands and saw the mockingbird species on each island differed slightly. They were suited perfectly to survive in the environment they were in. A beak didn't grow longer because an animal strained for it to grow. Something else was at work, and that something else was selection. A longer beak, and the random mutation which provided for it, gave the bird an advantage over others. That genetic modification was gifted to its offspring. The animal itself did nothing to help the genetic modification to occur. It simply happened. And because of it, along with other slight modifications along the way that provided further benefits, a new species developed. Just as would a new strain of virus. Viruses evolve more quickly because of the vast number involved. Viruses have high mutation rates, and generations can pass in a short period of time. As a result, we are constantly fighting to keep ahead of them.

Darwin hypothesized that animals evolved due to a process he called natural selection. And then he tested it. He strengthened his hypothesis with each new test and observation. He tied all the facts together using his hypothesis as the string and offered it to everyone else to test as well. Scientists have been doing so for over 150 years. Not only biologists, but chemists, geologists, paleontologists, physicists, and pretty much every other scientific discipline you can name. They have all tested Darwin's theory, and it has held up. It has been strengthened by every test, and it has not once been proven to be wrong. Not once. Because of this, it long ago graduated from a hypothesis to a theory. It is a valid explanation for the fact of evolution.

Many of us have gotten used to using the word theory in our everyday lives. We see things happen, we witness market trends or stock market dips, and we might say something like, "I have a theory about that." Everyone has their own personal theory that they might apply to situations and observations. That's because the word *theory* has entered the popular culture. We don't say, or you would be hard-pressed to find anyone who says, that we have a hypothesis. We say "theory" because it's easy, it's short, and we've heard it used by scientists for centuries. Because we are accustomed to using it to explain anything and are often proven wrong, the word has lost its weight. When presented with Darwin's "theory" some people immediately feel justified in discounting it. Their theory about why the bedroom door seems to stick in the summer and not the winter was wrong, so Darwin could just as equally be wrong. They don't realize that what they should have said, to give the proper distinction, was that their hypothesis about the bedroom door was wrong. If they approached their hypothesis of why the door sticks correctly and subjected it to a battery of tests, they might then be able to call it a theory if those tests ended with the same results.

So can Darwin's theory be proven wrong? Of course. Theories are theories because they have stood up to science's inquisitive and skeptical eyes. A theory allows for science to make predictions and to test those predictions. If at some point, a new fact is presented that doesn't support Darwin's theory, then science will pounce on it and attempt to explain why. British scientist J. B. S. Haldane is said to have famously stated that he would be convinced theories about evolution were wrong if a rabbit fossil was found in the Precambrian. That would mean it predated all other mammal fossils and was found in a level of rock strata that proved this to be the case.

To date, no such discovery has ever been made, and Darwin's theory of evolution by natural selection continues to be the best explanation for the world around us.

I think Darwin can rest his case.

Chapter Forty-Three

Evolutionary Psychology

One night, a very long time ago, millions of years ago, in fact, a young hominid heard a noise. There was a lot of noise at night. This one had sounded close. Perhaps too close. He sniffed around and suddenly froze.

He'd seen snakes before. His family always gave them a wide berth. This one was big. Bigger than any snake he had ever seen before. He didn't dare call out to his folks. The creature was looking right at him.

This young hominid's fear outlived him. His fear had been passed down to him, and he, in turn, passed it on to his children, and they to their children. I'm of course talking about a fear of snakes. Or pretty much anything that slithers low to the ground on its belly.

I grew up in New England. The snakes we usually run into are of the common garter variety. Nothing too terrifying. When I was a kid, I lifted a piece of old plywood in a field to find a milk snake curled up beneath it. I thought it was a coral snake and assumed it would bite me, so I ran. I didn't know coral snakes didn't live in my part of the country. Even today, when I unexpectedly run into a snake, my pulse quickens. I might pick it up now and move it to a safer location away from my lawnmower, but that uneasy feeling is still there. And I can't explain why.

The hominid at the beginning of this chapter could. If he were still alive, he might nod and tell me snakes are dangerous, and the knowledge of their wickedness has been passed down to me. Not verbally but genetically. My brain has been wired to fear snakes.

Is that true? In the 17th century, philosopher John Locke explained we were all born as a blank slate, a tabula rasa. Everything we learn, we learn by experience. We are not preprogrammed for anything. Least of all the fear of snakes.

What if we aren't born with an empty hard drive, but one bundled with a bunch of tiny little programs bundled along with it? Maybe these programs are coded to run at certain times so we can better navigate the world. Recognizing and understanding how these programs work could be the key to understanding how we work. Or, more specifically, our minds, behaviors, and fears.

Enter the discipline of evolutionary psychology.

Darwin pondered the idea that everything around us is a product of natural selection. Our attitudes and emotions are handed down to us by way of our internal wiring and not by experience. At least not completely. Experience has taught me not to run when I see a snake. But my "snakes are dangerous" program is still running. It might be outdated software, but it still serves a purpose should I ever encounter a coral snake.

I don't feel the same way when I get into a car or hold a gun. I might respect the power of the car's engine or the ability for me to do harm with a gun, but I'm not afraid when I encounter them. I encounter cars and guns a lot more frequently than I do snakes. I've known people who have been seriously injured by both. Just turn on the news, and you'll hear about a car accident or someone who's been shot. It would make a lot more sense for my pulse to quicken when I get into a car than when I see a snake.

So what's the story here? That's what evolutionary psychology seeks to explain. The question it asks is why we feel the way we do in certain situations. What psychological adaptations were naturally selected to accompany us on our journey forward through time?

We have eyes sensitive to shades of light, ears stimulated by vibrations in the air, and nerves on our skin to feel the world around us. These are adaptive traits we evolved with. Natural selection saw to that. Every living organism on this planet, be it plant or animal, was molded into the shape it is for a reason. Not a conscious, designed reason, but a reason dictated by the organism's ability to survive and reproduce. If an adaptation or mutation wasn't beneficial, then the organism disappeared.

When I jump at a sudden shadow or feel a sense of dread when I'm alone in the woods at night, those feelings may also be present because of an adaptive trait. Or, at the very least, my fears might be linked to genes that were selected to remain a part of me.

It doesn't stop at fears, either. Evolutionary psychology wants to know why people cheat on their spouses, why we lie, and why we stop to help someone who is hurt or in danger. These are all open for explanation and interpretation.

Your mind is not much different than a computer. It receives data input, analyzes it, and provides a response. Different parts of your brain handle different bits of sensory data. They have different functions. The physical configuration of your brain is the product of natural selection. The cognitive processes going on within that configuration are part of that selection process. The genes responsible for its physical structure were selected because they allowed for our survival. Our brains contain a history of physical *and* psychological adaptations. Just like an archaeologist digs into the sands of time to piece together the physical world, it may be possible to do the same for the psychological world.

Our society abhors incest. But why? I'm sure you can come up with many reasons, but other than saying it's gross or unnatural, what is the real reason? Why are we psychologically averse to the idea? That aversion is part of us. We may try to put it into words, but the fact of the matter is we all have this aversion. We weren't taught it. Our culture didn't tell us it was bad. It's just there—like laughter. Laughter evolved with us too. Our ancient ancestors used an early form of laughter to indicate they weren't aggressive. It helped to avoid conflict. Baby chimps laugh. Orangutans laugh. Even mice laugh. It is all part of the psychological makeup that was there when we were born. Locke's blank slate isn't so blank after all. Our minds are handed to us loaded with basic and essential software. Everything else, our wants, desires, and certain reactions are programs we picked up along the way. Our differences can be attributed to the different programs we've installed from the moment of birth to right now.

Traditional psychology asks how it is we function the way we do. Evolutionary psychology asks why. It looks at the brain as an information processing device. One which was shaped by natural selection to handle inputs and outputs the way it does. The neural mechanisms and pathways activated at any given moment are solving problems. Some of the solutions to those problems evolved with us. These are the very same problems our ancient ancestors faced. They don't have to be big problems either. When you have an itch on your nose, you scratch it. You don't consciously think about it. You raise your hand, and your fingers attend to the irritation. There is a complicated set of processes that are activated to perform this act, and not one of them are you consciously aware of or directing. You're not thinking about how to raise your hand or how to scratch the itch. Unconscious processes do this for you. You are aware of it but only milliseconds *after* the fact.

And what about the evolution of consciousness? How far back into our evolutionary past can we trace consciousness? Does it extend as far back as our reptilian ancestors? Are we still running some of the same unconscious programs for social interactions they did? We've inherited certain structures of our brain from them. Contained within those structures are stimulus-and-response programs that still work now just as they did then, like our amygdala. That's been with us for a very long time. It handles our flight or fight response. When I see a snake, I will retreat. When someone comes at me from a dark alley, I will defend myself. I don't think about these things. I react. And the source of that reaction, the shortcut to the response which avoids having to think about it, lies in an ancient program. A program that could be two million years old.

Our brains are physical systems. Could it really be that our responses, decisions, and attitudes are all directly tied to a physical set of neural circuits and nothing more?

It's not exactly a pleasant thought, is it? That our minds, our thoughts, and our reactions are all a set of programs that were installed two million years ago. Along the way, these programs have been changed depending on the environment we found ourselves in.

It goes back to the age-old question of nature vs. nurture. How much of what we do is the product of nature, or natural selection, instead of the culture we are born into, with all of its social nuances and rules? It's very likely a mixture of both. The percentage in the mix is up for debate. Dual Inheritance Theory, or DIT, was developed to help define this mixture. According to DIT, our behavior is a mix of genetic and cultural evolution. It's the perfect middle ground. It's up to evolutionary psychologists to decide where they fall on the playing field, with DIT as the dividing line.

When we think about animal actions, we often attribute them to instinct. Birds fly south in the winter. They don't think about it; they just do it. It's a migratory reflex. They follow an ancient program that works in tandem with their biology and the earth's magnetic fields. Like software and a computer.

But apply this same logic to our actions, and most people will cry foul.

Critics of evolutionary psychology have pointed out several things they find wrong with its theories and refuse to classify it as a science. For one, it's hard to test. I opened the chapter with a story about an ancient hominid and his fear of snakes—a fear that was inherent in his genes, his internal programming, and passed down to us. I presented it as a seemingly irrational fear with a seemingly rational explanation. But how do I test that? Or, to be more specific, from a scientific point of view, how can my theory of the hominid and his fear-snake gene be falsified? It's all speculation on my part, or so it would seem.

This leads us to evolutionary psychologists' biggest stumbling block: the criticism that their theories and conclusions are all "just so" stories, like Rudyard Kipling's book of that title, published in 1902, in which he fantasized about why camels have humps or how a leopard got its spots.

Perhaps all of this talk about linking behavior to genes is simply scientific justification based on speculation. The brain is an incredibly adaptive and flexible organ. There are many brain injury cases where one part is damaged only to have another part rewire itself to accommodate, or fill in, for the damaged area and functions affected. If certain areas of our brains are tied to certain behavioral traits due to natural selection, then how is it that those traits can be passed on to different areas of the brain under extreme circumstances or injuries? Could John Locke be right that it is not genetic programming that accounts for our behavior but instead your own experience? Is it that natural selection provides the hard drive, an empty one fresh out of the factory, and it's then up to us to install as many programs as it can hold?

Our hard drives are full of programs that run beneath the surface. We are unaware of them, but they are there. They help us to interact with and navigate the world. Without them, we'd have a tough time discerning a dangerous situation from a friendly one. There is also space set aside for new experiences. Things our ancient genes never accounted for. Like planes, trains, and automobiles, or how to greet someone from a culture completely different from our own. Those experiences become part of our own personal programs. They are not passed down, and they are unique to who we are until the power runs out.

Behind my home sits a very small two-hundred-year-old cemetery. I keep the weeds at bay and make sure the stones are clear of fallen branches after a bad storm passes through. It's a part of the past, and I like to take care of it. They were here long before I was born. The cemetery sits no more than thirty feet from my back deck. Every once in a while, late at night, when I'm writing or reading, I'll catch movement out of the corner of my eye. I look, and there's nothing there. Is it a ghost, a trick of the light, or a hyperactive agent detection device gifted to me by my genes?

I guess it depends on if you're a fan of scary stories.

Chapter Forty-Four

The Predictive Power of Evolution

What if I were to hand you a telescope that allowed you to view a world light-years away, a world similar to ours but covered with water? What if I told you intelligent beings live on that planet and explained they are more intelligent than we are and highly advanced? What do you think they look like? You might contemplate the fact that the planet appears to be 100 percent covered by water and conclude that it's populated by mermaids and mermen. You might speculate that their crafts look like our submarines. I might even agree with you.

But that's all we can do. Speculate. We can make broad predictive strokes when it comes to how an organism will evolve. What those changes will look like, if they happen at all, is beyond our power to know. We have more luck predicting when the next snowstorm will hit my little town in New Hampshire. Which is no luck at all.

Does this mean theories about evolution are outside of the realm of actual science?

A strong theory is testable. After analyzing a set of results, you can make predictions. When I mix blue and yellow liquids, the result will be a green liquid. I can vary the shade of green by varying the mixture. I can add red, and the color will be different. I can then say whenever I add blue and yellow together, I will get green. Blue and yellow being my primary colors and green the secondary color. I can tell you about this and how to test it. Knowing it all and reviewing my work, you can be pretty sure when you mix the two, following my very complicated instructions, the result will be green.

It's testable *and* predictive. I predict green, test it, and celebrate the results. We can confirm it or falsify it. It's scientific gold.

But what about evolution by natural selection? It's a scientific theory. But can I test it or make a prediction? I suppose I can try, but what sort of prediction would I make? What factors can I tweak?

The point is, we can't predict what will happen to an organism by tweaking environmental variables. We can't predict how it will evolve. Even if we were to observe its evolutionary path, it's highly unlikely we could replicate that test or be assured the result would always be the same. To be honest, it probably won't be.

This has been one of the criticisms that evolution research has faced. Philosopher Karl Popper is famous for saying evolutionary biology lacked any predictive power and was therefore untestable. He recanted his stance later in life, but in his statement is a bit of truth.

Or is there? Is it possible to raise the theory of evolution above the claim that it's all a matter of conjecture?

The answer, of course, is yes.

To better understand the predictive power of evolution, we must first chew on what it means to formulate a theory of prediction. We can then get to the real meat of the matter.

Let's take physics.

Physics fascinates me. We know so much about the world on a macro level that we can predict with confidence what will happen given a set of initial conditions and variables. It's how we can send ships to the moon or triangulate your coordinates so you can navigate from here to New York with the GPS app on your phone. Without the predictive power of science and repeatable tests, we would be able to do none of these things. Certain things happen consistently. We can bang flint stones together and get fire. Well, some people can. I've never mastered that.

If our efforts with the flint produced snowflakes instead of fire for no apparent reason at all, we wouldn't have made it very far as a species. We might still be hiding in caves and waiting for the sun to rise to hunt for raw food. We depend on the world around us to follow certain rules just as much as we seek to understand those rules. Prediction is power.

That's on the macro level. We can also perform tests and make predictions on the microscopic level. But push past into the quantum realm, where subatomic particles reign, and our predictive powers fall apart. In fact, we see many strange and unexplainable things. For now, at least. Theoretical physicist Richard Feynman is credited with saying if you think you understand quantum mechanics, you don't. Physics at the quantum level is too strange. Particles appear to pop in and out of existence. They can even become entangled where one particle seems to be tied to another particle no matter how far apart they're separated. Particles behave like waves and waves like particles. At the quantum level, determinism appears to be false. Prediction flies out the window.

The inability to predict a particle's path at the quantum level doesn't stop science from studying it and attempting to test it. We can make predictions about what *might* happen under certain conditions. That's because we can observe what *has* happened.

Let's look at this another way. I mentioned the weather and snowstorms. If you live in New England, you know how unpredictable the weather is during the winter months. It's almost ridiculous to try. It's a source of much frustration when planning a trip, even one to the grocery store. This past winter, there were massive snowstorms predicted which never appeared. Some of those predictions were for the very next day. One storm, the weather models predicted, was sure to hit us, so we prepared for it. I had the generator gassed up and ready to go. We expected school to be canceled, and I settled in to work from home. We went to bed expecting the worst and woke up the next morning to blue skies. The storm, overnight, had veered north of us. Maine was slammed while we went on with our normal daily lives. Sorry, Maine. My in-laws live there, and they were hit by the storm we were promised.

We can't predict what atmospheric conditions will be like with any level of certainty. Especially the further out we try to predict. There are simply too many variables and situations that will affect the outcome. But, the morning after the much-promised storm, when I woke up to blue skies and turned on the news, I could watch the weather videos and listen to the meteorologists interpret the data. I could then understand what happened and why it happened. It made perfect scientific sense. The meteorologists can explain it because they've seen it before and they understand it. After many years of observation and data collecting, they understood why the storm missed us. I might not, which is why I turn to them. Does that make meteorological science less powerful because it can't predict weather patterns with 100 percent certainty? Of course not. They study what has happened and apply that understanding to what is happening and what may happen. After the fact, they use the knowledge gained to better their understanding.

Would you like another example? How about forensics? We can't predict when a crime will happen. That's saved for science fiction. But once it has happened, and the culprit has left the scene, it's time for forensics to step in and piece together what occurred. Using science, they can tell you where someone stood when the fatal shot was fired, how they entered the room, how tall they might be, and what they wore if they left any fibers behind. As with meteorology, a forensic scientist pieces together the past.

So what's the weather going to be like a week from today? Nobody knows. Even the world's most powerful computer couldn't tell you. We can't feed all the variables into it that it would need to calculate such a thing. But given enough data from the past, it can begin to decipher patterns that occurred in the past. Using these patterns, meteorologists can make retrodictions. A retrodiction is simply a prediction set in the past.

Confused? Don't be. Let's step back into the world of evolutionary science.

Retrodictions rule when it comes to understanding how organisms have evolved. Let's look at dolphins. We can plot three points in time. We'll call these three points A, B, and C. A is located in the distant past. B is the more recent past. And C is 10,000 years from now.

Using the fossil record and genetics, we can see how dolphins evolved from point A to point B. We can make retrodictions as to the evolutionary path their ancestors took to get from A to B. But C, where they will travel, and what changes they will have undergone 10,000 years from now, we cannot do.

Knowing where an organism started on the tree of life, and knowing where it is now, allows us to uncover branches we didn't know existed. Having the right information makes retrodictions possible.

The problem is evolution moves slowly.

I didn't make up the word 'retrodiction' to prove my point. I found the first mention of it in a book published by the New York Academy of Sciences in 1877.

Charles Darwin knew the power of retrodiction, even if he didn't have a word for it. While studying the possible development of humans, he predicted that our ancient ancestors most likely walked out of Africa. We now believe this to be true. Both fossil and genetic evidence supports this theory. Then there are the Neanderthals. A favorite subject and mystery of mine. What happened to them? One theory proposes we gradually assimilated them. Again, genetic evidence is in support of that theory. DNA studies show that some of us, depending on our lineage, have very small genetic traces of Neanderthal DNA present within us.

But the best example of the predictive, or retrodictive, power of evolutionary science comes with the example of Professor Neil Shubin's discovery in 2004 of *Tiktaalik*, which we learned about in chapter 29. To me, this is the granddaddy of all examples. *Tiktaalik* is the fossilized remains of the half-fish, half-amphibian creature that lived 375 million years ago. It represents the transition of fish with gills and fins to land-dwelling tetrapods with lungs and legs. Sometime between 380 to 365 million years ago, this transition happened. Professor Shubin calls *Tiktaalik* a "fishapod."

And here's what makes Professor Shubin's discovery so amazing and the perfect example of the predictive power of evolution. He didn't find the fossil due to sheer blind luck, as so many people sometimes do. Professor Shubin knew about the fossils of fish from point A, which was 380 million years ago. He also knew about the fossils of land-dwelling tetrapods found at point C, 365 million years ago. So somewhere, at point B, existed a creature in between the two. If the fish-like animals at point A evolved to become the tetrapods at point C, using the theory of evolution by natural selection as a tool, he knew what those organisms at point B might look like. He didn't pick a random spot and start to dig for it. Instead, he figured out where he might find the fossil of a creature similar to what he was looking for. He predicted, and accurately, that the best place to find his fossil would be the Canadian arctic. The rocks were the right age.

Now that's science at its best. He could have dug anywhere, but using his knowledge of geology and how species have evolved, he was able to find exactly what he was looking for in the exact spot he expected to find it.

I'll give you one more quick example. This one, at least the prediction itself, predates Professor's Shubin's discovery by about 150 years. It was made by Charles Darwin while he was studying plants in 1862. There wasn't much Darwin didn't study. From barnacles to chimpanzees, Darwin was fascinated by the tree of life. One of the things that captivated him was the way flowering plants reproduced using pollen. There are quite a few methods for getting the pollen from one plant to another. In some cases, the wind is involved. In others, the plants have evolved flowers to attract insects.

A classic example of this is bees. Bees will land on a flower looking for nectar, inadvertently cover themselves with pollen, and travel to the next flower, spreading the pollen. The plants use bees to pollinate. The bees benefit from the flower, and the flower benefits from the bees. It's a match made in heaven. The flowers evolved to use bees in the pollen delivery process, and bees evolved to use the nectar.

What fascinated Darwin was a species of orchid from Madagascar that had an extremely long nectary. This is the part of the flower that produces nectar, the sugary liquid that some insects love. The orchid Darwin looked at had a nectary about a foot long. This puzzled him at first. The nectar collected at the bottom of this nectary. He realized the flower had evolved this way to protect its nectar from being collected by a random animal not involved in the pollination process. But what insect, he wondered, could ever reach it? In Darwin's own words, and this is a quote, "What insect can suck it?"

He predicted it had to be a moth with an extremely long tongue. The problem was, no known moth had a tongue that long. Twenty years after Darwin died, the moth he predicted was found. A large moth with a tongue a foot long. Did this moth actually use this tongue to get the nectar from Darwin's orchid? That question remained unanswered for almost a century. We know now that it does.

What sort of creatures will this planet be populated with in fifty thousand years? No one can say. Just like we can't predict what the weather will be like or the state of technology. That is, if there's anyone left to enjoy that technology. If things go the way we think they will, based on the past, we can say that the planet will most likely still be here and that life on this planet will continue to evolve. Mankind will evolve as well. Whether we are looking out at the universe with two organic eyes or enhanced cybernetic eyes, the chances are that what we see will be beautiful.

Chapter Forty-Five

We *Are* Unique

When I was young, I'd lay in bed and listen to the music coming from the stereo in the living room. One of my earliest memories includes the song "Fame" by David Bowie. I could go on and on about the songs that accompanied me through my childhood. The list is endless.

In a previous chapter, I asked the question, "Are we unique?" and then set about showing why we are not by looking at the animal kingdom. From tool use to altruism, it appears we are not as special as we might think.

But there are a couple of instances where the degree of separation between us and those we share the planet with is too great to ignore. It would be like standing on the shore of the Atlantic Ocean and trying to see Europe.

One of those instances, as you've probably guessed, is music.

You might be thinking we aren't the only animal gifted with music. Birds make music, as do whales. You might even say an elephant blows its own sort of trumpet.

Then you have "Piano Concerto No. 21: Andante" by Wolfgang Amadeus Mozart. Mozart was born in 1756, and the music he left us with transcends both space and time. He composed over 600 works, and each one carries with it a hint of something. Something that moves us. It's a glimpse of the spark that animated us 3.5 billion years ago. That spark originated in the Big Bang.

I could have used the music of the Beatles to illustrate this. I'm fond of using their work as an example of music that will survive as long as there are ears to hear it and devices to play it with. That's the one thing that makes me nervous about our digital age. When it comes to music, books, and the wealth of knowledge we have accumulated and translated into digitized code, when the lights all go out, it all goes with it.

But as long as we have vocal cords and air to breathe, there will be music.

As Homo sapiens, we first appeared on the scene some 200,000 years ago, give or take a few thousand years. We left Africa somewhere around 60,000 years ago, and with us spread music. In 2012 fragments of flute carved from animal bone were discovered in the Geißenklösterle cave in Germany. The fragments are dated at 43,000 years old. Needless to say, we have been making music for a very long time. In 1972, Professor Anne Draffkorn Kilmer from the University of California managed to translate what might be the oldest song ever committed to paper. In this instance, the paper is a set of clay tablets discovered in 1950. The tablets are 3,400 years old and contain, literally etched in stone, symbols written in cuneiform, an ancient Sumerian script.

Music has a powerful effect on us. Music can make you smile, shed tears, or fall in love. It can even urge you to sit back, look at the stars, and wonder what it is all about.

Space is infinite. That vast canvas of darkness above has always fascinated us. Before we had fire to gather around for warmth and protection, the night meant vulnerability. It sometimes meant death. Predators hunted us at night just as we pursued them during the day. One day we are animated, moving stealthily through long grass in search of food, and the next day we are the food. How were we to explain death? Why did we even feel the need to explain it?

The unknown captivated us and troubled us. Just as the night's mysteries drew us in, so did the desire to understand how it all worked. Perhaps there were beings like ourselves, beings that we couldn't see but were more powerful, who were responsible for it all. If we were to please them, they might help us out. They might provide for us. They might comfort us.

Out of all of earth's creatures, we seem to be the only ones cursed with the ability to ask "why?" Philosophers gazed up at the heavens and contemplated our place in the universe. At the same time, religious leaders and country pastors sought to explain the gaps. We are not the only animal to be troubled by death on a deep level. Chimpanzees mourn when a loved one dies, and elephants have rituals associated with death. In fact, elephants' intelligence led Aristotle to remark that they surpass all others in wit and mind.

We alone appear to have the ability to look back into the past to explain the present and prepare for the future. Is it, as Darwin said, only a matter of degree, or is it something more? Our ability to look into and dissect the smallest forms of matter to understand the building blocks of the universe might set us above all others. We build large colliders to smash invisible particles together, just as we once did with rocks to watch sparks. Be they from flint or quarks, the sense of awe those sparks arouse in us will never quite go away.

In form, we are unique. We have hands with opposable thumbs to shape the world around us as we seek to keep the predators at bay and to protect those we love. We transform our thoughts into symbols on paper to allow others, even those we have never met, a glimpse into our thoughts. The air is full of our radio transmissions, and some of those transmissions, even the earliest ones, are still out there. Traveling outward across the vast expanse of space.

If our uniqueness is one of degree, then perhaps there are others out there a few degrees more advanced than we are. They need not look like us or think like us, they don't even need to be made out of the same stuff as us, but we'd like to meet them just the same.

Chapter Forty-Six

An Endless Tale

The universe has expanded and continues to expand to this very day. It has done so for almost 14 billion years. Its expansion will one day leave us without neighboring galaxies to contemplate. It will be a vast blackness, and we will feel truly alone. But there's no need to worry because at this moment we live in a unique and fortunate time. While the universe has evolved since that first inexplicable "bang," life here on our humble planet has done so too. 3.5 billion years ago, an organism first wiggled in a primordial pond. It split, replicated, combined, swapped genetic material, clumped into groups, formed new shapes, emerged from the sea, and struggled to survive. These new multicellular organisms suffered mutations during replication that either helped them cope with the environments they found themselves in or faded into history. Some left traces while others didn't. Mary Anning, who had a knack for spotting and pulling fossils out of the cliffs near her home, helped us to understand what those traces meant. They provide us with a glimpse into the past while we endeavor to understand our own origins.

Our brains have evolved to make sense of it all. They no longer function as primitive programs meant to direct us to find food, shelter, and safety. They are now complex networks that allow us to consider ourselves and our reflections. Those reflections may be in a mirror, the still surface of a pond, or the animals above, below, and beside us. We have reached a point where we can peer into our genetic makeup, poke at our DNA, and consider ways to improve the time we have left as a species. We can look up at the stars while they are still there and consider how we fit in. In them, we see new environments and new challenges. In four billion years, our tiny sun will breathe its last gasp of helium and die. It may feel as if we have plenty of time to figure it all out, and we do. As I said, we are fortunate. Imagine if life started a few billion years later than it had, and we opened our eyes to realize we have only a few thousand years left or less.

As it stands, barring an unfortunate cataclysm or encounter with a rogue asteroid, we can rest assured the secrets are there for our wonderfully complex minds to unlock. You need not be a scientist to appreciate the depth of those secrets. All you need to do is sit in your backyard or visit the closest nature preserve should you not have a backyard (bring a chair). Look at the grass, the trees, the birds that flitter from branch to branch, and the squirrels scurrying back and forth for acorns. Ask yourself if it would be too bold to consider all life came from one single filament, as Erasmus Darwin did so many years ago. You can breathe a sigh of relief, for others have considered it for you and have followed the path of that filament through time. It leads to you, and to the squirrel whose tail nervously twitches at your presence, as well as the acorn she holds.

The tree of life has so many branches that you would have a hard time counting them all. Think of it instead like a long meandering river. It turns, splits off, and those splits separate again. All of those side tributaries can be traced back to the original river and back to its source. There are also tributaries and streams that eventually dry up, never to be seen again.

Suffice to say, everything you see, touch, hear, taste, and smell is part of the same story. It's an old story. The oldest one ever told.

Bibliography

Alleyne, Richard. "Scientist Craig Venter Creates Life for First Time in Laboratory Sparking Debate about 'Playing God.'" *The Telegraph*, May 20, 2010, www.telegraph.co.uk/news/science/7745868/Scientist-Craig-Venter-creates-life-for-first-time-in-laboratory-sparking-debate-about-playing-god.html.

Aristotle. *History of Animals*. Translated by Arthur L. Peck. Cambridge, MA: Harvard University Press, 2001.

———. *Posterior Analytics*. Translated by Harold Percy Cooke. Cambridge, MA: Harvard University Press, 1997.

Big Think. "Michio Kaku: Mankind Has Stopped Evolving." YouTube, uploaded by Big Think, May 31, 2011, www.youtube.com/watch?v=UkuCtIko798.

The Cambridge History of Science. 8 vols. Cambridge: Cambridge University Press, 2018.

Carroll, Sean B. *Endless Forms Most Beautiful: The New Science of Evo Devo and the Making of the Animal Kingdom*. W. W. Norton, 2005.

Chambers, Robert. *Vestiges of the Natural History of Creation Together with Explanations: A Sequel*. Cambridge: Cambridge University Press, 2011.

Charter for Compassion. "Robert Wright: The Evolution of Compassion." YouTube, uploaded by Charter for Compassion, 15 May 2012, www.youtube.com/watch?v=mEDtyYwwJcU.

Coyne, Jerry A. *Why Evolution Is True*. New York: Viking, 2010.

CSHL DNA Learning Center. "Accumulating DNA Mutations through Time, Mark Stoneking." Video interview. Cold Spring Harbor Laboratory DNA Learning Center. www.dnalc.org/view/15168-Accumulating-DNA-mutations-through-time-Mark-Stoneking.html.

Darwin, Charles. *The Autobiography of Charles Darwin*. New York: Barnes and Noble Books, 2005.

———. *The Descent of Man, and Selection in Relation to Sex*. New York: Penguin Classics, 2011.

———. "Notebooks on Geology, Transmutation, Metaphysical Enquiries and Reading Lists." Darwin Online, darwin-online.org.uk/EditorialIntroductions/vanWyhe_notebooks.html.

———. *On the Origin of the Species by Natural Selection of the Preservation of Favoured Races in the Struggle for Life*. New York: Signet Classic, 2003.

Darwin, Erasmus. *The Temple of Nature; or, the Origin of Society*. The Project Gutenberg e-Book. www.gutenberg.org/files/26861/26861-h/26861-h.htm.

———. *The Temple of Nature*. Canto I. knarf.english.upenn.edu/Darwin/temple1.html.

Dawkins, Richard. *The Blind Watchmaker: Why the Evidence of Evolution Reveals a Universe without Design*. New York: W.W. Norton, 2015.

———. *The Greatest Show on Earth the Evidence for Evolution*. New York: Free Press, 2009.

———. "Richard Dawkins on Altruism and the Selfish Gene." YouTube, uploaded by Andy80o, September 1, 2012, www.youtube.com/watch?v=n8C-ntwUpzM.

Dickens Journals Online. "Mary Anning, the Fossil-Finder.," www.djo.org.uk/indexes/articles/mary-anning-the-fossil-finder.html. Accessed November 6, 2020.

Diderot, Denis. *Rameau's Nephew and D'Alembert's Dream*. New York: Penguin Books, 1966.

———. *Thoughts on the Interpretation of Nature and Other Philosophical Works*. Translated by Lorna Sandler. Manchester, UK: Clinamen, 2000.

"E. O. Wilson (04/03/12)." Charlie Rose. YouTube, April 4, 2012, www.youtube.com/watch?v=j4Ltmy4DvNg.

Geological Society of London. Letter from Mary Anning, [1833]. https://stage.geolsoc.org.uk/Library-and-Information-Services/Exhibitions/Women-and-Geology/Mary-Anning/Letter-from-Mary-Anning. Accessed December 27, 2021.

Goodall, Jane. "What Separates Us from Chimpanzees?" YouTube, TED, May 16, 2007, www.youtube.com/watch?v=51z7WRDjOjM.

Gould, Stephen Jay. *Wonderful Life: The Burgess Shale and the Nature of History*. New York: W.W. Norton, 2007.

Greenwood, Veronique. "A Horse Has 5 Toes, and Then It Doesn't." *New York Times*, February 10, 2020. www.nytimes.com/2020/02/08/science/horses-toes-hooves.html.

Hutton, James. *Theory of the Earth; or an Investigation of the Laws Observable in the Composition, Dissolution, and Restoration of Land upon the Globe*. archive.org/stream/cbarchive_106252_theoryoftheearthoraninvestigat1788/theoryoftheearthoraninvestigat1788#page/n1/mode/2up.

Lamarck on Use and Disuse. www.ucl.ac.uk/taxome/jim/Mim/lamarck6.html.

Lehrer, Jonah. "Kin and Kind." *The New Yorker*, March 5, 2012. www.newyorker.com/magazine/2012/03/05/kin-kind.

Lucretius. *The Nature of Things*. Translated by A. E. Stallings. New York: Penguin Books, 2015.

Maillet, Benoît de. *Telliamed; or, Conversations between an Indian Philosopher and a French Missionary on the Diminution of the Sea*. Edited by Albert V. Carozzi. Urbana: University of Illinois Press, 1968.

Matthew, Patrick. *On Naval Timber and Arboriculture*. London: Longman, 1831.

Maupertuis, Pierre Lois. *Venus Physique* (The Earthly Venus). cogweb.ucla.edu/EarlyModern/Maupertuis_1745.html.

Miller, Kenneth R. *Only a Theory: Evolution and the Battle for America's Soul*. New York: Penguin Books, 2009.

Paley, William. *Natural Theology; or, Evidences of the Existence and Attributes of the Deity Collected from the Appearances of Nature*. London: J. Faulder, 1810.

Prothero, Donald R., and Carl Dennis Buell. *Evolution: What the Fossils Say and Why It Matters*. New York: Columbia University Press, 2007.

Richard Dawkins Foundation for Reason and Science. "Richard Dawkins: Comparing the Human and Chimpanzee Genomes—Nebraska Vignettes #3." YouTube, uploaded by Richard Dawkins Foundation for Reason & Science, 5 June 2014, www.youtube.com/watch?v=WMPlr4tD64A.

Ruse, Michael, and Joseph Travis. *Evolution: The First Four Billion Years*. Cambridge, MA: Harvard University Press, 2011.

"Self-Recognition in Apes National Geographic." YouTube, uploaded by National Geographic, March 13, 2008. www.youtube.com/watch?v=vJFo3trMuD8.

Shermer, Michael. *Why Darwin Matters: The Case against Intelligent Design*. New York: Holt, 2007.

Shubin, Neil. *Your Inner Fish a Journey into the 3.5-Billion-Year History of the Human Body*. New York: Vintage Books, 2009.

Stott, Rebecca. *Darwin's Ghosts: The Secret History of Evolution*. New York: Spiegel and Grau, 2012.

"Theory." *Merriam-Webster Dictionary*. Merriam-Webster, www.merriam-webster.com/dictionary/theory.

"Time Magazine Interviews: Dr. Jane Goodall." YouTube, uploaded by *Time*, September 16, 2009, www.youtube.com/watch?v=t7iIT7fZFZ8.

University of California Museum of Paleontology. "Mary Anning (1799–1847)." ucmp.berkeley.edu/history/anning.html. Accessed November 6, 2020.

Wallace, Alfred Russel. *My Life: A Record of Events and Opinions*: 1823–1913. London: Chapman and Hall, 1970. Internet Archive, archive.org/details/b31360580_0001.

Waterfield, Robin, and David Bostock. *Aristotle Physics*. Oxford: Oxford University Press, 1996.

Watson, James. "How I Discovered DNA." YouTube. TED-Ed, 26 July 2013, https://www.youtube.com/watch?v=RvdxGDJogtA.

Web of Stories. "John Maynard Smith: The Idea of Sexual Selection (30/102)." YouTube, September 6, 2011, www.youtube.com/watch?v=h4kkn0l8BZk&feature=youtu.be.

"William Charles Wells." Wikipedia. April 2020, en.wikipedia.org/wiki/William_Charles_Wells.

Wilson, Edward O. *The Meaning of Human Existence*. New York: Liveright, 2015.

Yale Courses. "20. Coevolution." YouTube, uploaded by Yale Courses, September 1, 2009, www.youtube.com/watch?v=fUKWpF2sK34.